FOREWORD

The importance of Environmental N
any organization can never be over(
product has its manufacturer's manual. Man has limited knowledge
of natural processes which are slow and go unnoticed. Due to human
operational activities, these natural processes are aggravated,
resulting in unprecedented negative impact and devastating
consequences on the environment and human life. In order to
mitigate the negative impacts of such human operational/business
activities, EMS has become a ready tool for organizational
performance evaluation as well as process, product and services
evaluation. Organizational commitment and respect for the
environment has now become a tool for product evaluation and
added value for customer patronage and organizational credibility.

EMS is a must for every organization, no matter how small.
Organizations, Professionals and Students lack simple text books on
EMS, especially in developing countries where vulnerability is very
high. In consequence of this gap, this book is expected to go a long
way in contributing positively to improving the knowledge and skills
in EMS within any business management platform.

Emmanuel O. Emmanuel (PhD)
July 2014

TABLE OF CONTENTS

LIST OF TABLES

LIST OF FIGURES

CHAPTER ONE

1.0 INTRODUCTION

1.1 What Is Environmental Management System (EMS)?

Environmental Management System (EMS) can be described as a program of continuous environmental improvement that follows a defined sequence of steps drawn from established project management practice and routinely applied in business management. In simple terms these steps are as follows:

• Review the environmental consequences of the operations.
• Define a set of policies and objectives for environmental performance.
• Establish an action plan to achieve the objectives.
• Monitor performance against these objectives.
• Report the results appropriately.
• Review the system and the outcomes and strive for continuous improvement.

Not every system will present these steps in exactly the same way, but the basic principles are clear and easily understandable.

Environmental Management System (EMS) is a management approach, which enables an organization to identify, monitor and control its energy and environmental aspects. An environmental management system is part of the overall management system that includes organizational structure, planning activities, responsibilities, practices, procedures, processes and resources for developing, implementing, achieving, reviewing and maintaining the energy and environmental policy. Specifically, an EMS ensures that energy and environmental programs and operations are enhanced through a commitment to continual improvement, a focus on aligning processes and procedures with goals and objectives, and a clear definition of responsibilities for energy and environmental issues from the top to the bottom of the organization.

EMS such as ISO 14000 series are seen as mechanisms for achieving improvements in environmental performance and for supporting the trade prospects of "clean" firms. The potential advantages of EMS are rewarding, especially the emphasis on performance improvement and for simplification of certification, as well as the potential for regulatory streamlining; and the trade consequences.

1.2 The Development of EMS Standards

The world's first standard for environmental management systems (EMS) was BS 7750. It was developed and published by the British Standards Institute (BSI) in 1992. This standard was the model for the ISO 14000 series developed by the International Organization for Standardization (ISO). ISO 14001, which establishes the requirements for an EMS, was finalized in 1996. BS 7750 was also the basis for the European Union's Eco-Management and Audit Scheme (EMAS).

Because ISO14001 and EMAS are both based on BS7750, all the three standards are quite similar in their approach. If an organization complies with BS7750, little effort will be needed to fulfill the requirements of ISO14001 or EMAS. The ISO 14001 standard for EMS is a voluntary standard for environmental management.

1.3 Elements of EMS

A basic environmental management system typically includes the elements presented below. The level of detail of an environmental management system varies between organizations, depending on the nature of operations (for example, housing, water provision, and sanitary systems), or other activities that are closely related to the environment and natural resources (for example, forestry, agriculture, aquaculture, mining, oil and gas).

Basic Environmental Management System Environmental policy

• Environmental statement/vision

• Environmental objectives and indicators Implementation strategy or

• Strategy to attain the goals set out in the policy and to environmental action plan integrate environmental considerations in all aspects of decision-making and all stages of the initiative's cycle

• Environmental roles and responsibilities (within the organization and between the organization and its partners) and financial resources

• Environmental awareness and training programs

• Follow-up and evaluation mechanisms to ensure that the implementation strategy is effective

• Assessment of lessons learned (using results-based management) and method of re-integrating these lessons into the planning and/or performance monitoring process of initiatives.

Associated tools:

• Environmental guidelines and/or criteria for the selection and monitoring of initiatives (Guidelines and standards for environmental auditing).

• Environmental assessment procedure in accordance with the criteria to be applied early in the planning process, through a participatory approach and including a mechanism to capture "projects" requiring the completion of such an environmental assessment, as well as a reporting and filing process (Guidelines and procedures for environmental impact assessment, EIA).

• Other tools to assist the environmental assessment procedures or to facilitate the integration of environmental considerations (e.g. checklists)

1.4 The Benefits of an EMS

An environmental management system (EMS) is a structured program of continuous environmental improvement that follows procedures drawn from established business management practices. The concept is straightforward, and the principles can be easily applied, given the necessary support. There has been increasing interest in the potential value of EMS approaches, of which the recently released ISO 14000 series is the most widely known. The first steps in the control of industrial pollution have been the creation of the necessary regulatory framework and the specification and design of control equipment to reduce emissions.

These efforts have been broadly successful in improving the performance of many polluters, but in other cases, investments in pollution equipment are wasted because the equipment is not operated properly. Attention, the World Bank and elsewhere, is turning to support regulatory and end-of-pipe approaches through incentives, production efficiencies, and management improvements—a range of measures often grouped under the broad banner of cleaner production and eco-efficiency. The potential benefits of eco-efficiency are unequivocal: good operational practices, supported by committed management, can achieve considerable improvements in environmental performance at low cost and can get the maximum benefits from investments in hardware. Without management and worker support, the best equipment can be useless. The challenge is to achieve long-lasting improvements in performance, and EMS is seen as one of the key tools in achieving this.

An important related issue, in a context of increasingly free trade, is the concern that environmental performance may become an important commercial factor, either as a positive attribute or as a potential trade barrier. The implementation of an EMS, and particularly of the ISO 14000 system, is seen as a way to demonstrate an acceptable level of environmental commitment. A good EMS allows an enterprise to understand and track its environmental performance. It provides a framework for implementing improvements that may be desirable for financial or other corporate reasons or that may be required to meet regulatory requirements. Ideally, it is built on an existing quality management system, referred to as ISO 9000 series and deals with Quality Assurance.

1.5 Environmental and economic benefits of ISO

ISO has a multi-faceted approach to meeting the needs of all stakeholders from business, industry, governmental authorities and nongovernmental organizations, as well as consumers, in the field of the environment.

1. ISO has developed Standards that help organizations to take a proactive approach to managing environmental issues: the ISO 14000 family of environmental management standards which can be implemented in any type of organization in either public or private sectors – from companies to administrations to public utilities.

2. ISO is helping to meet the challenge of climate change with standards for greenhouse gas accounting, verification and emissions trading, and for measuring the carbon footprint of products.

3. ISO develops normative documents to facilitate the fusion of business and environmental goals by encouraging the inclusion of environmental aspects in product design.

4. ISO offers a wide-ranging portfolio of standards for sampling and test methods to deal with specific environmental challenges. It has developed some 570 International Standards for the monitoring of such aspects as the quality of air, water and the soil, as well as noise, radiation, and for controlling the transport of dangerous goods. They also serve in a number of countries as the technical basis for environmental regulations.

CHAPTER TWO

2.0 ISO, ENVIRONMENT AND OTHER STANDARDS

2.1 ISO and the environment

ISO International Standards and related normative documents provide consumers, regulators and organizations in both public and private sectors with environmental tools with the following characteristics :

- Technically credible as ISO standards represent the sum of knowledge of a broad pool of international expertise and stakeholders
- Fulfill stakeholder needs as the ISO standards development process is based on international input and consensus
- Facilitate the development of uniform requirements as the ISO standards development process is built on participation by its national member institutes from all regions of the world
- Promote efficiencies when the same standards are implemented across markets, sectors, and/or jurisdictions
- Support regulatory compliance when the standards are used to meet market and regulatory needs
- Enhance investor confidence because the standards can be used for conformity assessment such as by audit, inspection or certification. This enhances confidence in products, services and systems that can be demonstrated to conform to ISO standards and provides practical support for regulation.

Organizations around the world, as well as their stakeholders, are becoming increasingly aware of the need for environmental management, socially responsible behavior, and sustainable growth and development. Accordingly, as the proactive management of environmental aspects converges with enterprise risk management, corporate governance, and sound operational and financial practices and performance, international standards are becoming increasingly important for organizations to work towards common and comparable environmental management practices to support the sustainability of their organizations, products, and services.

Furthermore, governments and regulatory bodies are increasingly looking to ISO standards to provide a framework to ensure alignment and consistency both nationally and internationally.

2.2 ISO 14000 and Other Standards

Presently, there are two major areas in the evaluation of environmental management practices. One area focuses on organizational issues, while the other focuses on the products, services and processes. The ISO 14000 series covers the following topics:

1. Organization Evaluation.
    -   Environmental Management Systems (ISO 14001, 14004)
    -   Environmental Performance Evaluation (ISO 14014, 14015, 14031)
    -   Environmental Auditing (ISO 14010, 14011, 14012, 14013, 14014)

2. Products, Services and Processes
    -   Life Cycle Assessment (ISO 14040, 14041, 14042, 14043)
    -   Environmental Labeling (ISO 14020, 14021, 14022, 14023, 1402X)
    -   Environmental Aspects in Product Standards (ISO 14060)

For a better understanding of EMS standards, one should note that *EMS standards are process standards and not performance standards*. These include organization process standards and the product and services standards. In other words, the EMS standards do not tell organizations what environmental performance they must achieve (besides compliance with environmental regulations). Instead the standards describe a system that will help an organization achieve its own set objectives and targets. *The assumption is that better environmental management will lead indirectly to a better environmental performance.* The good thing about this approach is that environmental users are indirectly compelled to becoming environmental managers. In the past, environmental management had been domiciled in the hands of government without positive results or progress. Every person is an environmental user, but every person is not an environmental manager. With effective environmental management, all environmental users should be involved in the process of sustaining the environment we depend on. In summary, the management of the environment is the responsibility of all environmental users, especially organizations and institutions (See Figure 2.1).

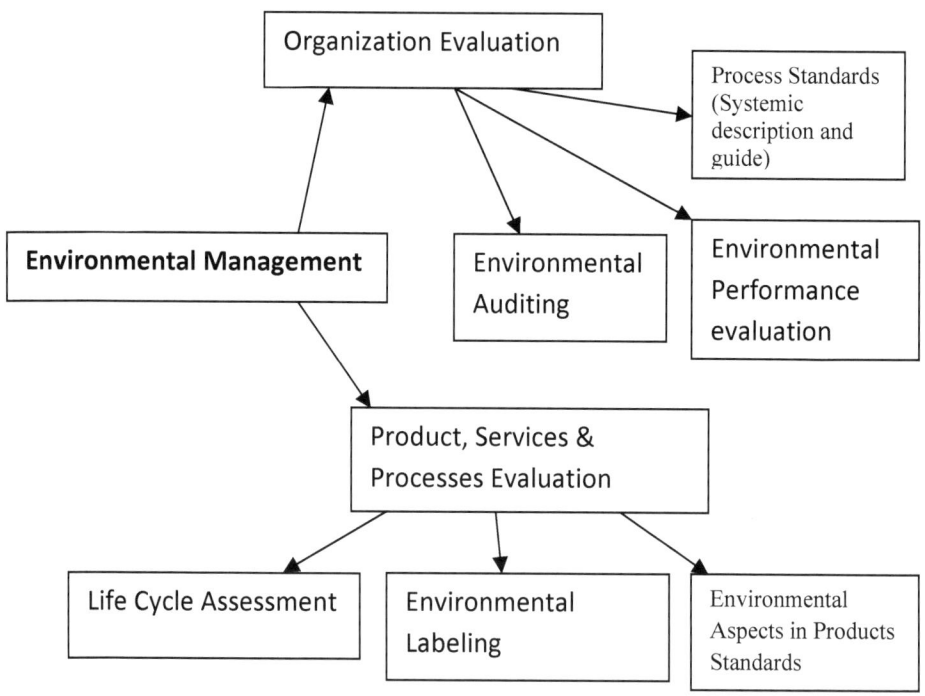

FIG 2.1: ENVIRONMENTAL MANAGEMENT

If an EMS were adopted purely as an internal management tool, the details of the system and its structure would not be important. However, the EMS is becoming more and more a matter of interest to people outside the management of the enterprise—to workers, regulators, local residents, commercial partners, bankers and insurers, and the general public. In this context, the EMS is no longer an internal system and becomes a mechanism for communicating the enterprise's performance to outside parties, and some level of standardization and common understanding is required.

The best-known common framework for EMS is the ISO 14000 series. This series is based on the overall approach and broad success of the quality management standards prepared and issued as the ISO 9000 series. ISO 14000 consists of a series of standards covering eco-labeling and life cycle assessment (LCA), as well as EMS. The documents formally adopted (by the end of 1996) as international standards are those covering EMS: ISO 14001 and ISO 14004. There are two other major EMS standards: the British BS 7750, which was one of the first broadly accepted systems and has been adopted by a number of other countries, and the European Eco-Management and Audit Scheme (EMAS). A process of harmonization has been under way to ensure reciprocal acceptability of these systems with ISO 14001. BS 7750 and EMAS are, however, broader in their requirements than ISO 14000. In particular, EMAS includes requirements for continued improvement of performance and for communication with the public, which are not part of ISO 14001. Within the ISO system, ISO 14001 sets out the basic structure for an EMS, while ISO 14004 provides guidance. The crucial feature of the ISO 14001 standard is that it identifies the elements of a system which can be independently audited and certified. The issue of certification underlies much of the discussion about environmental management systems. The presentation in these standards is clear and concise and provides a framework that can be used as the starting point for a simple system for a small company or a highly detailed one for a multinational enterprise. Compliance with ISO

14001 does not by itself automatically ensure that an enterprise will actually achieve improved environmental performance. *The standard requires that there be an environmental policy that "includes a commitment to continual improvement and pollution prevention" and "a commitment to comply with relevant environmental legislation and regulations". It also requires that the enterprise establish procedures for taking corrective and preventive action in cases of nonconformance.* The desirable approach would be for management to make a commitment to specific environment performance improvements within a defined period and then use ISO 14000 as the mechanism for demonstrating that it is complying with that commitment. Management commitment includes budgetary allocation for EMS and the appointment of a Management Representative.

It should be noted that ISO 14000 standards are voluntary. Organizations that embrace EMS are encouraged on continual basis to sustain their improved environmental performance and are not dragged to court for non-conformance. "Adoption" by a country normally means that the national standards organization has said that the ISO version is the EMS standard that is recognized. It does not imply any formal requirement that companies adopt such a standard.

CHAPTER THREE

3.0 ENVIRONMENTAL MANAGEMENT AND SUSTAINABLE DEVELOPMENT

3.1 Environmental management and sustainability

ISO technical committee ISO/TC 207, Environmental management, is responsible for developing and maintaining the ISO 14000 family of standards. The committee's current portfolio consists of 21 published International Standards and other types of normative document, with another nine new or revised documents in preparation.

ISO/TC 207 was established in 1993, as a result of ISO's commitment to respond to the complex challenge of "sustainable development" articulated at the 1992 United Nations Conference on Environment and Development in Rio de Janeiro. It also stemmed from an intensive consultation process, carried out within the framework of the ISO Strategic Advisory Group on Environment (SAGE). SAGE was established in 1991 and brought together representatives of a variety of countries and international organizations – a total of more than 100 environmental experts – who helped to define how International Standards could support better environmental management.

As a result, the ISO 14000 family of standards for environmental management was launched to provide a practical toolbox to assist in the implementation of actions supportive to sustainable development.

From its beginning, it was recognized that ISO/TC 207 should cooperate closely with ISO/TC 176, Quality management and quality assurance – the ISO technical committee responsible for the ISO 9000 family of quality management standards – in the areas of management systems, auditing and related terminology.

Successful steps have been taken to ensure compatibility of the ISO 14001 and ISO 9001 standards to facilitate their use by organizations that wish to implement both environmental and quality management systems to benefit themselves and their customers and stakeholders.

These steps include a common standard (ISO 19011) giving guidelines for auditing environmental and/or quality management systems.

3.2 Global participation
Membership of ISO/TC 207 is among the highest of any ISO technical committee and is both broad and diverse in representation, two key indicators of the worldwide interest in the work of this technical committee. National delegations of environmental experts from over 100 countries participate in ISO/TC 207, including 27 developing countries. The leadership of the committee is " twinned " between a developed and developing country (currently Canada and Brazil).

3.3 ISO/TC 207, Environmental management
3.3.1 Origins Compatibility
The national delegations are chosen by the national standards institute concerned and they are required to bring to ISO/TC 207 a national consensus on issues being addressed by the technical committee. This national consensus is derived from a process of consultation with interested parties and stakeholders in each country. ISO/TC 207 continues to explore new and innovative ways to allow member countries to contribute and participate in the standards development process without increasing their carbon footprint. ISO/TC 207 has relationships with over 30 international organizations that serve as liaison members to the technical committee.
These organizations include the following:
- Asian Productivity Organization
- Confederation of European Paper Industries
- European Commission
- Environmental Defense Fund
- Global Eco-labeling Network
- International Aluminum Institute
- International Chamber of Commerce
- International Institute for Sustainable Development

- International Iron and Steel Institute
- Organization for Economic Co-operation and Development
- Sierra Club
- United Nations Environment Program
- World Business Council for Sustainable Development
- World Health Organization
- World Resources Institute
- World Trade Organization.

Published documents and ongoing work of ISO/TC 207 address the following areas:
- Environmental management systems
- Environmental auditing and related environmental investigations
- Environmental performance evaluation
- Environmental labeling
- Life cycle assessment
- Environmental communication
- Environmental aspects of product design and development
- Environmental aspects in product standards
- Terms and definitions
- Greenhouse gas management and related activities
- Measuring the carbon footprint of products.

The ISO 14000 family of standards reflects international consensus on good environmental and business practice that can be applied by organizations all over the world in their specific context.

3.3.2 Scope of ISO/TC 207's work

ISO 14001 is the world's most recognized framework for environmental management systems (EMS) that helps organizations both to manage better the impact of their activities on the environment and to demonstrate sound environmental management. ISO 14001 has been adopted as a national standard by more than half of the 160 national members of ISO and its use is encouraged by governments around the world. Although certification of conformity to the standard is not a requirement of ISO 14001, at the end of 2007, at least 154 572 certificates had been issued in 148 countries and economies. ISO 14001 addresses not only the environmental aspects of an organization's processes, but also those of its products and services. Therefore ISO/TC 207 has developed additional tools to assist in addressing such of products and services from the "cradle to the grave" : from the extraction of resource inputs to the eventual disposal of the product or its waste.

Other environmental management tools developed by ISO/TC 207 include the following:

ISO 14004 complements ISO 14001 by providing additional guidance and useful explanations.
Environmental audits are important tools for assessing whether an EMS is properly implemented and maintained. The auditing standard, ISO 19011, is equally useful for EMS and quality management system audits. It provides guidance on principles of auditing, managing audit programs, the conduct of audits and on the competence of auditors.

ISO 14005 provides guidelines for the phased implementation of an EMS to facilitate the take-up of EMS by small and medium-sized enterprises. It will include the use of environmental performance evaluation.

ISO 14006 provides guidelines on eco-design.

ISO 14010 provides guidelines for environmental auditing: general principles of environmental auditing

ISO 14011 provides guidelines for environmental auditing: audit procedures; auditing of environmental management systems.

ISO 14012 provides guidelines for environmental auditing: qualification criteria for environmental auditors.

ISO 14015 provides guidelines for Environmental Assessment of Sites and Organizations

ISO 14020 series of standards addresses a range of different approaches to environmental labels and declarations, including eco-labels (seals of approval), self-declared environmental claims, and quantified environmental information about products and services.

ISO 14021 provides guidelines for Environmental Labels and Declarations - Self-declared Environmental Claims

ISO 14024 provides guidelines for Environmental Labels and Declarations - Type I Environmental Labeling - Guiding Principles and Procedures

ISO 14025 provides guidelines for Environmental Labels and Declaratives - Type III Environmental Declaratives - Principles and Procedures

ISO 14031 provides guidance on how an organization can evaluate its environmental performance. The standard also addresses the selection of suitable performance indicators, so that performance can be assessed against criteria set by management. This information can be used as a basis for internal and external reporting on environmental performance.
Communication on the environmental aspects of products and services is an important way to use market forces to influence environmental improvement. Truthful and accurate information provides the basis on which consumers can make informed purchasing decisions.

ISO 14033 provides guidelines and examples for compiling and communicating quantitative environmental information.

ISO 14040 standards give guidelines on the principles and conduct of Life Cycle Assessment studies that provide an organization with information on how to reduce the overall environmental impact of its products and services (Principles and Procedures).

ISO 14044 is on Environmental Management – Life Cycle Assessment – Requirements and Guidelines

ISO 14047 is on Environmental Management - Life Cycle Assessment - Examples of Application of ISO 14042

ISO 14048 is on Environmental Management - Life Cycle Assessment - Data Documentation Format

ISO 14049 is on Environmental Management - Life Cycle Assessment - Examples of Application of ISO 14041 to Goal and Scope Definitions and Inventory Analysis

ISO 14050 is on Environmental Management - Vocabulary

3.4 Overview of the ISO 14000 family of standards

3.4.1 Published standards
ISO Guide 64 provides guidance for addressing environmental aspects in product standards. Although primarily aimed at standards developers, its guidance is also useful for designers and manufacturers.

The ISO 14000 series is still under development. The Technical Committee, TC 207 is supervising and developing the ISO 14000series. The Secretariat of TC 207 is held by the Standards Council of Canada (SCC). Each standard undergoes different stages, namely,

- Preliminary Work Item (PWI)
- New Work Item (NWI)
- Working Draft (WD)
- Committee Draft (CD)
- Draft International Standard (DIS)

The standards can be classified into four (4) groups according to their focus:

1. Organization and systems standards
   - ISO 14001 Environmental management systems -- Requirements with Guidance for Use
   - ISO 14004 Environmental Management Systems -- General Guidelines on Principles, Systems and Supporting Techniques

2. Evaluation and auditing standards
   - ISO 14010 Guidelines for environmental auditing: general principles of environmental auditing
   - ISO 14011 Guidelines for environmental auditing: audit procedures; auditing of environmental management systems.
   - ISO 14012 Guidelines for environmental auditing: qualification criteria for environmental auditors.
   - ISO 14013 Management of Environmental Audit Programs
   - ISO 14014 Initial Reviews
   - ISO 14015 Environmental Site Assessment
   - ISO 14031 Evaluation of Environmental Performance

3. Production-oriented standards
   - ISO 14020 Environmental Labels and Declaratives - General Principles
   - ISO 14021 Environmental Labels and Declarations - Self-declared Environmental Claims
   - ISO 14022 Environmental Labels and Declarations - Symbols
   - ISO 14023 Environmental Labels and Declarations – Testing and Verification Methodologies
   - ISO 14024 Environmental Labels and Declarations - Type 1 Environmental Labeling - Guiding Principles and Procedures
   - ISO 1402X Environmental Labels and Declarations – Type 111 Labeling
   - ISO 14040 Environmental Management - Life Cycle Assessment - Principles and Procedures

- ISO 14041 Environmental Management - Life Cycle Assessment - Life Cycle Inventory Analysis
- ISO 14042 Environmental Management - Life Cycle Assessment – Impact Assessment
- ISO 14043 Environmental Management - Life Cycle Assessment -- Interpretation
- ISO 14060 Guide for the inclusion of Environmental Aspects in Product Standards

4. Definition standards
   - ISO 14050 Environmental Management – Vocabulary—Guide to the Principles for ISO/TC 207/SC Terminology Work.

3.4.2 New standards

Sustainable development policy and practice has attracted considerable attention and debate in the past 15 years. Our understanding of and concerns about environmental and sustainable development issues have also evolved over time. Just as the existing ISO 14000 standards play an important role in helping organizations to address today's priorities, so too can future standards help to address future priorities.

An integral part of an organization's EMS is the commitment to continual improvement. ISO/TC 207 takes this principle to heart and is constantly improving its process to identify and respond to new standardization needs.

ISO/TC 207's success in continuing to work on relevant standards is evidenced by the development of the following new standards:

ISO 14045 will provide principles and requirements for eco-efficiency assessment. Eco-efficiency relates environmental performance to value created. The standard will establish an internationally standardized methodological framework for eco-efficiency assessment, thus supporting a comprehensive, understandable and transparent presentation of eco-efficiency measures.

ISO 14051 will provide guidelines for general principles and framework of material flow cost accounting (MFCA). MFCA is a management tool to promote effective resource utilization, mainly in manufacturing and distribution processes, in order to reduce the relative consumption of resources and material costs. MFCA measures the flow and stock of materials and energy within an organization based on physical unit (weight, capacity, volume and so on) and evaluates them according to manufacturing costs, a factor which is generally overlooked by conventional cost accounting. MFCA is one of the major tools of environmental management accounting (EMA) and is oriented to internal use within an organization.

ISO 14063, on environmental communication guidelines and examples, helps companies to make the important link to external stakeholders.

ISO 14064 parts 1, 2 and 3 are international greenhouse gas (GHG) accounting and verification standards which provide a set of clear and verifiable requirements to support organizations and proponents of GHG emission reduction projects.

ISO 14065 complements ISO 14064 by specifying requirements to accredit or recognize organizational bodies that undertake GHG validation or verification using ISO 14064 or other relevant standards or specifications.

ISO 14066 will specify competency requirements for greenhouse gas validators and verifiers.
The development program of ISO/TC 207 is constantly evolving, driven by market needs. Hence the above is a small sample of areas where standards are currently in development. Please consult www.iso.org for an up-to-date program of standards under development by ISO/TC 207.

ISO 14067 on the carbon footprint of products will provide requirements for the quantification and communication of greenhouse gases (GHGs) associated with products. The purpose of each part will be to: quantify the carbon footprint (Part 1) ; and harmonize methodologies for communicating the carbon footprint information and also provide guidance for this communication (Part 2).

ISO 14069 will provide guidance for organizations to calculate the carbon footprint of their products, services and supply chain.

Although the ISO 14000 standards are designed to be mutually supportive, they can also be used independently of each other to achieve environmental goals. The whole ISO 14000 family of standards provides management tools for organizations to manage their environmental aspects and assess their environmental performance. Together, these tools can provide significant tangible economic benefits, including the following:

- Reduced raw material/resource use
- Reduced energy consumption
- Improved process efficiency
- Reduced waste generation and disposal costs
- Utilization of recoverable resources.

Of course, associated with each of these economic benefits are distinct environmental benefits too. This is the contribution that the ISO 14000 series makes to the environmental and economic components of sustainable development and the triple bottom line. The standards that have been adopted are shown in Table 3.1.

TABLE 3.1: Overview of the ISO 14000 family of standards.

ISO 14000 Series	Environmental Management Systems
ISO 14001: 2004	Environmental management systems -- Requirements with Guidance for Use
ISO 14004: 2004	Environmental Management Systems -- General Guidelines on Principles, Systems and Supporting Techniques
ISO 14005:	Environmental Management Systems -- Guidelines

(2010)	for a Staged Implementation of an Environmental Management System, including the Use of Environmental Performance Evaluation.
ISO 19011: 2002	Guidelines for Quality and/or Environmental Management Systems Auditing
ISO Q19011S: 2004	Guidelines for Quality and/or Environmental Management Systems Auditing: US Version with Supplemental Guidance Added
ISO 14010-1996	Guidelines for environmental auditing: general principles of environmental auditing
ISO 14011-1996	Guidelines for environmental auditing: audit procedures; auditing of environmental management systems.
ISO 14012-1996	Guidelines for environmental auditing: qualification criteria for environmental auditors.
ISO 14015: 2001	Environmental Assessment of Sites and Organizations
ISO 14020: 2000	Environmental Labels and Declaratives - General Principles
ISO 14021: 1999	Environmental Labels and Declarations - Self-declared Environmental Claims
ISO 14024: 1999	Environmental Labels and Declarations - Type I Environmental Labeling - Guiding Principles and Procedures
ISO 14025: 2006	Environmental Labels and Declaratives - Type III Environmental Declaratives - Principles and Procedures
ISO 14031: 1999	Environmental Management - Environmental Performance Evaluation - Guidelines
ISO 14032: 1999	Environmental management - Examples of Environmental Performance Evaluations
ISO 14040: 2006	Environmental Management - Life Cycle Assessment - Principles and Procedures

ISO 14044: 2006	Environmental Management – Life Cycle Assessment – Requirements and Guidelines
ISO 14047:2003	Environmental Management - Life Cycle Assessment - Examples of Application of ISO 14042
ISO 14048: 2002	Environmental Management - Life Cycle Assessment - Data Documentation Format
ISO 14049: 2000	Environmental Management - Life Cycle Assessment - Examples of Application of ISO 14041 to Goal and Scope Definitions and Inventory Analysis
ISO 14050: 2009	Environmental Management - Vocabulary
ISO 14062: 2002	Environmental Management - Integrating Environmental Aspects into Product Design and Development
ISO 14063: 2006	Environmental Management - Environmental Communications - Guidelines and Examples
ISO 14064-1: 2006	Greenhouse Gases - Part 1: Specification with Guidance at the Organizational Level for Quantification and Reporting of Greenhouse Gas Emissions and Removals
ISO 14064-2: 2006	Greenhouse Gases - Part 2: Specification with Guidance at the Project Level for Quantification, Monitoring, and Reporting of Greenhouse Gas Emission Reductions or Removal Enhancements
ISO 14064-3: 2006	Greenhouse Gases - Part 3: Specification with Guidance for Validation and Verification of Greenhouse Gas Assertions
ISO 14065: 2007	Greenhouse Gases -- Requirements for Greenhouse Gas Validation and Verification Bodies for Use in Accreditation or other Forms of Recognition

3.4.3 The ISO 14000 family and the PDCA cycle

The ISO 14000 family is designed to be implemented according to the same Plan-Do-Check-Act (PDCA) cycle underlying all ISO management systems standards. The following table classifies the standards making up the ISO 14000 family according to their optimal place in the PDCA cycle.

TABLE 3.2: Plan-Do-Check-Act cycle.

PLAN	DO	CHECK	ACT
Environmental management system implementation	Conduct life cycle assessment and manage environmental aspects	Conduct audits and evaluate environmental performance	Communicate and use environmental declarations and claims
ISO 14050:2009 Environmental management – Vocabulary	**ISO 14040:2006** Environmental management – Life cycle assessment – Principles and framework	**ISO 14015:2001** Environmental management – Environmental assessment of sites and organizations (EASO)	**ISO 14020:2000** Environmental labels and declarations – General principles
ISO 14001:2004 Environmental management systems – Requirements with guidance for use	**ISO 14044:2006** Environmental management – Life cycle assessment – Requirements and guidelines	**ISO 14031:1999** Environmental management – Environmental performance evaluation – Guidelines	**ISO 14021:1999** Environmental labels and declarations – Self-declared environmental claims (Type II environmental labelling)
ISO 14004:2004 Environmental management systems – General	**ISO/TR 14047:2003** Environmental management – Life cycle impact	**ISO 19011:2002** Guidelines for quality and/or environmental management	**ISO 14024:1999** Environmental labels and declarations –

guidelines on principles, systems and support techniques	assessment – Examples of application of ISO 14042	systems auditing	Type I environmental labelling – Principles and procedures
ISO/DIS 14005 Environmental management systems – Guidelines for the phased implementation of an environmental management system, including the use of environmental performance evaluation	**ISO/TS 14048:2002** Environmental management – Life cycle assessment – Data documentation format		**ISO 14025:2006** Environmental labels and declarations – Type III environmental declarations – Principles and procedures
			ISO/AWI 14033 Environmental management – Quantitative environmental information – Guidelines and examples
Address environmental aspects in products and product standards		Evaluate greenhouse gas performance	

ISO Guide 64:2008	ISO/TR 14049:2000	ISO 14064-3:2006	ISO 14063:2006
Guide for addressing environmental issues in product standards	Environmental management – Life cycle assessment – Examples of application of ISO 14041 to goal and scope definition and inventory analysis	Greenhouse gases – Part 3 : Specification with guidance for the validation and verification of greenhouse gas assertions	Environmental management – Environmental communication – Guidelines and examples
ISO/CD 14006 Environmental management systems – Guidelines on ecodesign	**ISO/CD 14051** Environmental management – Material flow cost accounting – General principles and framework	**ISO 14065:2007** Greenhouse gases – Requirements for greenhouse gas validation and verification bodies for use in accreditation or other forms of recognition	
	ISO/WD 14045 Eco-efficiency assessment – Principles and requirements		
	Manage greenhouse gases		
ISO/TR 14062:2002 Environmental	**ISO 14064-1:2006** Greenhouse	**ISO/CD 14066** Greenhouse	

management – Integrating environmental aspects into product design and development	gases – Part 1: Specification with guidance at the organization level for quantification and reporting of greenhouse gas emissions and removals	gases – Competency requirements for greenhouse gas validators and verifiers document	
	ISO 14064-2:2006 Greenhouse gases – Part 2 : Specification with guidance at the project level for quantification, monitoring and reporting of greenhouse gas emission reductions or removal enhancements		
	ISO/WD 14067-1 Carbon footprint of products – Part 1: Quantification **ISO/WD 14067-2**		

	Carbon footprint of products – Part 2: Communication		
	ISO/AWI 14069 GHG – Quantification and reporting of GHG emissions for organizations (Carbon footprint of organization) – Guidance for the application of ISO 14064-1		

CHAPTER FOUR

4.0 GENERAL APPLICATIONS

4.1 IMPLEMENTING AN ENVIRONMENTAL MANAGEMENT SYSTEM

4.1.1 Commitment and Environmental policy

An environmental policy is a statement of the organizations overall aims and principles of action with all relevant regulatory requirements. It is a key tool in communicating the environmental priorities of your organization to employees at all levels, as well as to external stakeholders. As such, it should be written clearly and concisely to enable a layperson understand it, and should be made publicly available. It is up to the organization to decide on environmental priorities based on an initial environmental review, but these choices should be justified in the policy. To be truly effective, the policy should regularly be reviewed and revised and incorporated into the organization's overall corporate policy. The policy statement should set in writing a few achievable quantifiable priorities related to the environmental management system and the significant environmental effects found at the work-site.

To ensure an organization's commitment towards a formulated environmental policy, it is essential that top management is involved in the process of formulating the policy and of setting priorities. Based on this commitment, the organization should then conduct an initial environmental review and draft an environmental policy. The approved environmental policy statement must be communicated internally and made available to the public.

The commitment includes three basic policy statements:

- Continuous improvement in environmental Performance
- Compliance with environmental regulations
- Maintaining public relations regarding environmental issues of the organization, its activities and services.

The format to the environmental policy takes into account the following elements

- Title of the environmental policy.
- The need for an environmental policy.
- Statement of policy (Commitment and environmental issues).
- Procedure and guidelines for implementing the environmental policy.
- Responsibility and authority.
- Definition of terms.

4.1.2   Initial Environmental Review

An initial environmental review covers all the aspects of an EMS. As a result of this review, the organization knows its strengths, weaknesses, risks and opportunities regarding the current status of its EMS. The gaps between the requirements of the EMS standard and the actual status of the organization shows which aspects the organization should focus its efforts on to improve the system.

This leads directly to the development of an environmental management program that should fill the gaps. The review should focus on three key areas:

- Examination of existing environmental management practices and procedures.

- Identification of significant environmental impacts and their priority.
- Identification of legal and regulatory requirements.

Once the review has been conducted, all the information should be presented in a report (environmental review report). The following can be adopted;

- **Title page**

    State the title clearly; naming the organization and activity, state the name of the author, his or her position, organization, address, telephone, fax and email. Include the date of submission, the name of the addressee of the report and the copyright.

- **Executive Summary**

    As the name implies, this is a brief Summary of the review results. Describe the background and context of the environmental review here and state upon whose request it was carried out. Also draw attention to the most outstanding findings and recommendations, calling for immediate action.

- **Scope of Review**

    In this section, give details of the issues investigated (environment management system, environmental impacts and environmental regulations).

- **Finding of Facts**

    Summarize the responses to the questionnaire the reviewer's impressions of conversations with personnel, observations of the premises and examination of the documents here.

- **Recommendations**

    Prioritize recommendations according to the need for immediate corrective action and long- term goals.

- **Conclusions**

Sum up the findings and recommendations into a positive statement of action.

- **Appendices**

    Include the sample questionnaire, summaries of interviewers and other important documents.

Based on the results of the initial environmental review, a project can be developed to build up an EMS or to adapt to the current EMS to comply with one of the EMS standards.

4.1.3 Planning the Environmental policy

An organization should formulate a plan to fulfill its environmental policy. The environmental management system elements related to planning include (ISO 14004):

- Identification of environmental aspects
- Evaluation of associated environmental impacts
- Legal requirements
- Environmental policy
- Internal performance criteria
- Environmental objectives and targets
- Environmental management programs

    The initial environmental review is the source of much of the data needed for the planning process. In consequence, it is important to conduct a review. The review will help to identify priority environmental aspects and provide information on how far the organization complies with legal and other requirements. This in turn helps in the formulation of objectives and targets, which are then used to develop an environmental management program.

The latter is basically a systematic, detailed means of achieving organizational objectives and targets. If the organization does not have an EMS, the first management program should be the implementation of an EMS.

The identification of significant environmental impacts is done by developing an environmental exposure portfolio for an organization, its activities, products and services. This evaluation of the importance of certain environmental problems to an organization and the analysis of the cause of these problems lead to a set of environmental aspects that are significant for an organization.

The identification of legal and regulatory requirements assesses two levels of an organization.

- Production-related environmental regulations
- Product and service-related environmental regulations

The former addresses the production department while the latter addresses the Marketing and Research and Development Departments. To obtain information about environmental regulations, the following information sources can be used:

- Government Authorities
- Industry Associations
- Internet
- University publications (Law Departments)

The objectives of the organization should be specific while the target should be measurable. Setting objectives requires an analysis of the exposure of an organization to different environmental aspects; this should be in line with the initial environmental review. Environmental objectives should be set for the following environmental aspects as derived from the exposure analysis.

- Environmental aspects which have high public priority and to which the organization contributes heavily. Here, environmental objectives should be set and given the highest priority because these environmental problems are very sensitive and are likely to have an impact on the organizations free cash flow.

- Environmental aspects which have low public priority and to which the organization contributes heavily. Here, environmental objectives should be set and given high priority because these environmental problems are sensitive and are likely to have an impact on the organizations free cash flow if public priority changes (e.g. due to new scientific knowledge etc.) The objectives should be to keep an eye on possible changes in public perception and hence, priorities, and to prepare alternatives.

- Environmental aspects which have high public priority and to which the organization has a low contribution. These objectives should be added to the above if any investments or change in technology (products and production processes) are planned. Because of the high public priority, the objective should be to hold the current position by not contributing more to these problems.

Later on, these objectives are used for evaluating environmental performance because performance can only be evaluated by comparing the current state to a target level or objective. It is of great importance that the targets are set on an environmental problem level rather than on a causal level. Example, if a company wants to reduce its global warming impact by 10%, reduction in fossil fuel consumption would be set as a target. This restricts the organizations option. But, if an approach based on the problem (impact) rather than the cause is followed, a company has at least four more options to achieve the target of reducing its impact on global warming.

i)   Reducing CFC emissions

ii)   Changing the mix of energy consumed (e.g. substituting natural gas for oil)

iii)   Installing a catalytic end-of-the-pipe methane reducing facility

iv)   Banning the use of HCFCs

All the above measures lead to a decrease in global warming. The question is which alternative has the least cost or the best return, while achieving the target. In other words, which alternative has the best efficiency assuming the effectiveness is comparable (10% reduction in the contribution to global warming).

Environmental Management program is a description of the means of achieving environmental objectives and targets. It gives you what must be done, by whom, how and when for each of the defined objectives and targets of high priority.

- Designing responsibility for achieving objectives and targets at each relevant function and level.
- Providing the means (human resource, skills, technology, and financial resource e.t.c) for fulfilling the objectives and targets.
- Designating a time frame within which objectives and targets will be achieved.

To be most effective, the environment management program should be integrated into the organizations strategic plan. The environmental management plan should be periodically reviewed and regularly revised to reflect changes in the organizations objectives and targets.

4.1.4 Implementing the environmental policy

An effective implementation of the environmental policy calls for an organization to develop the necessary capabilities and support mechanisms to achieve its environmental policy, objectives and targets. For many organizations, implementation can be approached in stages, depending on the level of awareness of environmental requirements, aspects, expectations and benefits and the availability of resources.

The following aspects are covered in the implementation stage:

i) **Structure and Responsibility**

The roles, responsibilities and authority of personnel whose activities have an impact on the environment are defined, documented and communicated to all members of the organization. In addition, the resources required for the implementation and maintenance of the EMS are provided.

A special management representative with the responsibility and authority to enforce the EMS requirements is appointed. This management representative reports the environmental performance of the organization directly to top management.

ii) **Training, Awareness and Competence**

Management must ensure that personnel throughout the organization are aware of the environmental policy, the environmental management programs and the actual or potential impact of their activities on the environment. All personnel with a significant contribution to environmental performance need to be adequately trained to handle the environmental aspects of their activities. The organization must ensure that these people have the competence to deal with their responsibilities either through education, training and/or experience. Training should be in-depth enough to enable employees to integrate relevant environmental aspects into their daily business information.

iii)   Internal Communication

Internal Communication mechanism should be established between various levels and functions of the organization. An organization can communicate in a variety of ways including bulletin, notice board postings, meetings, e-mail, e.t.c. Communication should be a two-way process and the information communicated should be understandable and adequately explained. There must be a feed-back mechanism to avoid misinterpretation or misconception. Communication is also needed to demonstrate management commitment, raise awareness, deal with concerns and questions about the organizations activities, products or services, and to inform interested parties about the organizations EMS and performance.

Results from EMS monitoring, audit and management review should be communicated to those within the organization who are responsible for performance.

iv)   Environmental management system Documentation

The EMS of an organization must be well-documented in order to enable external auditors certify the management system according to one of the current EMS standards. The documentation includes a description of the basic elements of the system and their interaction. It also points to related documents, which may include:

a) Process information
b) Organization Charts
c) Internal Standards and operational procedures
d) Site emergency plans

v)   **Document Control**

All documents associated with EMS in an organization should be revised, reviewed and approved regularly so that updated information is always available. Such documents should be under the control of the designated environmental management representative.

vi) **Operational Control**

To ensure that the environmental policy is implemented and objectives and targets are achieved, and organization must identify processes and activities that have a significant impact on the environment. The organization must ensure that these processes and activities are conducted as intended. Those include procedures to control operations such as documented work instructions to ensure conformance to the EMS, monitoring and control of relevant process characteristics (e.g. effluent waste to streams and waste disposal). The organization should also establish procedures for verification of compliance to specified requirements and establish and maintain records of the results. In addition, the responsibility and authority for measures to deal with non-compliance and subsequent corrective actions should be specified. The organization should also identify environmental aspects of the goods and services it uses and communicate its environmental requirements to its suppliers and contractors.

vii) **Emergency Preparedness and Response.**

The organization must identify the potential for accidents and emergency situations and develop appropriate procedures to respond to them, as well as the prevention and mitigation of their impact on the environment. These procedures should be communicated internally and tested to make sure that the response is effective and efficient.

The emergency plan can include the following:

- Details of possible accidents (e.g. Chemical Spills, Fire).
- Actions to be taken if accidents occur.
- List of key personnel and their responsibilities in handling emergency situations.
- Drills to test effectiveness.
- Regular training on safe conditions and safe acts.
- Routine check schedules on safe conditions and safe acts.

4.1.5 Measurement and Evaluation.

After implementing the environmental policy, management needs to measure environmental interventions and their impact on the environment. In order to achieve this, management needs to open an environmental effects register where the environmental inventory is recorded. All equipments used for monitoring and measuring must be accurate and calibrated on a regular basis.

The environmental data management involves the recording of physical environmental data, environmental regulations and environmentally induced financial information. Environmental data management is necessary as a basis for effective decision making. In other words, the ecological, legal and financial data systems must be built up from scratch or adapted to the requirements of the EMS standards.

Environmental effects register is a collection and preparation of physical environmental information. It is a relatively new taste for managers. To check the compliance status of an organization, additional information about regulations and other requirements is needed. The procedure for building up an environmental effects register follows the methodology of accounting, but all the figures are measured in; kg, t, kWh or other units of measurement.

Four steps are essential in the process:

1) Definition of the systems (periods of time selected, organizational units included). The period of time over which the data is recorded is essential. It is a common practice today to use the same time period as in financial accounting e.g., one year. This means that all environmental data is measured in units of input per year or unit of output per year. Apart from the period of time, there must be a clear definition of the organizational units to be included.

2) Classification of environmental accounts (Structures, names, units).

   Once the operational units are selected, the data to be collected must be defined. Special attention should be given to the structure of the accounts and the existing sources of environmentally relevant data, such as accounting practices for materials, the amount of energy used, site permits for particular pollutants, production statistics, technical specification of the production machines etc. To obtain a better picture of protection, the accounting system should reflect environmental costs. Information about environmentally induced costs and earning needs to be collected. All this information should be recorded in such a manner that the data can be verified by an internal or external auditor as exemplified in the format below.

TABLE 4.1: Input and Output Inventory.

Inputs	Yr 1 (Qtr)	Yr 2 (Qty)	Outputs	Yr 1 (Qty)	Yr 2 (Qty)
Raw Materials(t)			Products(millions) - Books - Magazines - Brochures - Calendars		
Main Operating materials (t)			Recyclable Wastes (t) - Paper - Wood		

			- Printing Plates		
Other Operating materials (t)			Non-recyclable Waste(t) - Coated Paper - Construction materials - membranes		
Hazardous materials (t) -Chemicals -Gasoline -Grease			Hazardous Waste (t) - Paints/pigments - Solvents - Filter towels - Developers - Oily and solid wastes		
Energy (Millions MJ) -Electricity -Natural gas -Gasoline -Propane			Water Emissions - Waste water - Effluent discharge		
			Air Emissions (t) - Carbon dioxide - Carbon monoxide - Sulfur dioxide - Nitrogen Oxide - Particulate matter - Methane - Non-methane volatiles - Organic compounds		

3) Selecting an appropriate basis.

The environmental effects register discussed above shows all the figures as inputs or outputs per year. This comparison does not take into account the effect of an increase or decrease in production (i.e. more input or more output). To demonstrate the success of an effective EMS, production fluctuations have to be eliminated.

The basis should correlate strongly to major;

- Inputs such as raw materials and/or energy
- Outputs such as waste or sewage.

Having selected an appropriate basis, the effectiveness of environmental management programs, in terms of reduced environmental impact per unit of output can be assessed.

4) Collection and recording of data.

Collection and recording of data starts after the system is clearly defined, the specific accounts for the organization have been set up and a choice has been made regarding the basis for a yearly comparison. Collecting data, recording relevant information connected to it is the final step in building up an environmental effects register. Here, special attention should be given to documentation. The following data structure format may be considered.

TABLE 4.2: Data Structure of Environmental Accounts.

Data Structure	Particulars
Account Number:	300
Account name:	Crude Oil
Specification of the Input or Output:	Low Sulfur Fuel
Year:	2012
Amount:	10000
Unit:	tons
Source of information:	Energy Statistics
Assumption and/or Calculations:	Density = 860kg/m3
Information collection date:	01/01/2012
Name of person in charge:	Mr. Energy

Accounting benefits from an environmental effect register because the environmentally induced costs, such as energy costs or pollution abatement costs, can be allocated to the cost centers and the cost drivers that cause them. It should be noted that effective information management requires careful consideration concerning appropriate software.

Environmental Regulations Register

Environmental regulations register is often installed and maintained for the purpose of providing additional information about regulations and other requirements needed in checking the compliance status of an organization. The identification of legal and regulatory requirements assesses two levels of an organization.

i)   Production-related environmental regulations
ii)   Product and services-related environmental regulations.

The former addresses the production department while the latter addresses the Marketing and Research and Development departments. Fundamentally, two questions need to be addressed, namely:

i)   What are the relevant environmental regulations? (−Target)
ii)   What is the present situation in the organization? (=Actual)

To obtain the information about environmental regulations, the following information sources can be used:

- Government Authorities
- Industry Associations
- Law Firms

- Universities (Law Departments)
- Libraries

Financial environmental data provides information about environmentally-induced cost and earnings. Corporate and public environmental protection can only be successful if it is economically sustainable. In Consequence, it is important to gather economic information in environmental protection. Different methods of accounting and information management have been in use and mutually beneficial structures and relationship of reporting have been established. In practice, accounting systems are often not differentiated according to environmentally-induced information. All cost, clean-up costs, emission reduction costs etc. This information gives a preliminary indication of what fraction of total costs is due to environmental issues and how much money can be saved by better environmental performance.

Environmental performance evaluation is to provide decision-makers with a transparent and rational methodology for making environmentally sound decisions, based on environmental objectives and targets and the data collected. The performance assessment refers to the methodology used during the initial environmental review.

Environmental performance needs to be assessed on the following three levels:

- Assessment of Environmental Effects
- Assessment of Legal Compliance
- Assessment of Eco-Efficiency.

In the Assessment of environmental effects, the data provided by the environmental effects register is analyzed to assess performance against environmental objectives and targets. Environmental objectives are defined according to different environmental problems. The contribution of the organization to these problems is assessed e.g. the different releases to the environment (carbon dioxide and methane emissions etc.) within broader categories (global warming). All releases to the environment are assessed according to their actual impact. The most widely used approach to assess environmental impact is the two-step approach of *classification* and *characterization.* In this approach, environmental interventions are clustered by their potential link to a specific environmental problem and then assessed according to their contribution to the problem. For example Methane is a gas that can be classified as an emission with an impact on global warming. When its relative impact is characterized, it can be seen that the greenhouse potential of 1kg of methane is about 20 times as high as the potential of the same amount of carbon dioxide (based on actual scientific knowledge). Therefore all CO_2 emissions (kg) are multiplied by 1, and all methane emissions (kg) are multiplied by 20. As a result, two figures are derived which add up to a so-called "global warming potential" (measured in terms of kg CO_2 equivalent).

The Characterization factors for evaluating contributions to global warming are shown in the table below:

TABLE 4.3: Characterization Factors for Evaluating Contribution to Global Warming.

Characterization Factors for Evaluating Contributions to Global warming. **Objective:** Global Warming **Indicator:** Global warming Potential (kg CO2 Equivalent)	
Intervention: 1kg of	Global Warming Potential
Carbon Dioxide (CO2)	1kg CO2 Equivalent
Methane (CH4)	21kg CO2 Equivalent
Dinitrogen Oxide (N2O)	290kg CO2 Equivalent

CFC - 11	3500kg CO2 Equivalent
HCFC (R 134 a)	1200kg CO2 Equivalent
Halons	5800kg CO2 Equivalent
Carbon Monoxide (CO)	11kg CO2 Equivalent
Nitrogen Oxide (NOx)	8kg CO2 Equivalent

For every environmental objective with high priority, an environmental performance figure should be calculated accordingly. As a result, the organization has a set of key figures that measure environmental performance based on the environmental effects register. These figures are compared to the objectives and targets. If the organization does not meet the objectives and targets that had been set by top management, corrective actions should be taken.

Assessment of Legal compliance requires that the organization should analyze the internal implementation of all the regulations applicable to the organization, its activities, products and services. Based on the environmental regulations register, two questions must be answered:

- Does the organization comply with regulation "x"?
- How does the organization comply with the requirements of regulation "x"?

The status of internal compliance can be classified into the following categories:

i)   Non-Compliance with major environmental regulations.
ii)   Non-Compliance with minor environmental regulations.
iii)   Non-Compliance, but expected to comply in the near future.
iv)   100% compliance with current environmental regulations.
v)   100% compliance with current and expected.

Status (i) to (iii) does not meet the requirements of ISO 14011, EMAS or BS7750. Here, immediate non-conformance and corrective actions should be taken. Status (iv) is the minimum requirement. From the assessment, the organization knows its current status regarding legal compliance as well as opportunities and threats arising from changes in its legal environment.

Assessment of Eco-Efficiency involves the analysis of environmental and economic performance and is the key component in sustainable business management. Environmental performance indicators should always be put into the context of economic performance. Return on net assets (RONA) or shareholder value (SV) is examples of adequate measures of the economic performance of a company. When environmental performance (vertical axis) is plotted against economic performance (horizontal axis), the result shows the quadrangle where the organization belongs. Four positions can be distinguished in an eco-efficiency portfolio (See Figure 4.1).

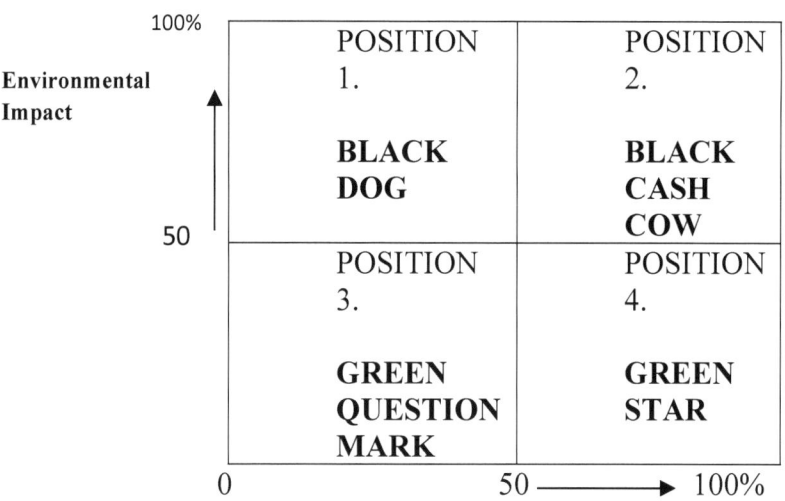

Economic performance

FIGURE 4.1: Eco-Efficiency Portfolio. (Source: Modified from Sturm, 1998)

The details are explained below:

1. Position 1 is Black Dog.

Organizations in Black Dog position have a high environmental impact added and have a weak economic performance. This position is economically weak and causes enormous environmental damage. Business in this situation should be eliminated or improved economically and ecologically. Changes in public environmental policy, like the implementation of green taxes will likely force such organization out of business.

2. Position 2 is Black Cash Cow.

Black Cash Cow position is characterized by relatively high financial revenues and a high environmental impact added. Assuming that governments will introduce a market based environmental policy mix and hence give a price tag to resource consumption and environmental impact, such organizations are likely to move to a 'Black Dog' position.

3. Position 3 is Green Question Mark.

These organizations have a low environmental impact added, but also achieve a relatively low economic performance. If an organization in this position cannot move to Green Star position, it cannot contribute to sustainable development as it cannot prevail because of its economically weak position.

4. Position 4 is Green Star.

A Green Star organization has low environmental impact added and a high economic performance. Low costs are achieved through integrated clean technologies. The environmental impact is already optimized when the technology was developed.

4.1.6 Audits and Review

This is essential components of EMS. Audits are used to determine if an organization is performing well with regards to its EMS, while review is used to get it back on track if it is not. Audits and reviews differ in the sense that audits are not meant to review the EMS itself, even though auditors may provide recommendations if a prior agreement has been made with the client to do so. It is top management's responsibility to determine if corrective actions should be taken (management review), using information obtained from the audit, as well as other sources. Audit can be defined as a systematic, documented, periodic and objective evaluation of the performance of an organization, management system and processes designed to protect the environment. The information gathered during an audit will be used for verifying the organization's management system and environmental statement. In consequence, it is important to allocate sufficient time and resources for an audit to achieve it s objectives. The main purpose of carrying out an audit is to help management control all the work practices which have an effect on the environment and to assess how far the organization is complying with its own environmental policy. Apart from the main objective of determining conformance to EMS criteria, an EMS audit can serve several other objectives:

- Determining whether the EMS is properly implemented and maintained
- Identifying areas for potential improvement of the EMS
- Assessing the capability of the internal management review process to ensure the continuing sustainability and effectiveness of the EMS

- Evaluating the EMS as part of a potential contractual relationship.

Audits can be carried out by both internal personnel and/or external parties selected by the organization. In either case, the auditors should be independent, impartial, objective and properly trained. As expertise is needed from various fields, audits are generally carried out by teams. The audit team should be independent of the site audited, as any external verifier would require that this condition be met. The audit should have a clearly defined scope, as defined by the lead auditor and the client/auditee (the person for whom the audit is being carried out). In the scope, describe the extent of the audit and set the boundaries (i.e. the site, which activities, how the results are reported). Before embarking on the audit, prepare a list of all the activities and subjects to be covered. To avoid confusion, make sure that any subsequent changes are agreed to by the lead auditor and the auditee. It is important to establish an audit cycle to ensure that audits of the most significant activities (e.g. treatment of effluents and hazardous wastes) are carried more frequently than activities that have little environmental effect. Report of audit should be presented to management once it is ready. The report should outline the scope of the audit and provide management with the information on the state of compliance with the organization's environmental policy and the environmental progress at the site. It should also include information on the effectiveness and reliability of the arrangements for monitoring environmental impacts at the site. While the responsibility for initiating corrective action is up to top management (determined after the review), the need for corrective action, where appropriate, should be demonstrated in the audit report.

Management review is carried out, based on the audit findings. Management should conduct a review to assess the continuing stability, adequacy and effectiveness of the EMS. The review should be broad enough in scope to address the environmental dimensions of all activities, products or services of the organization, including their impact on financial performance and possibly competitive position. Review of policies, objectives and procedures should be carried out by the level of management that defined them. The review process should include the following (ISO 14004):

Review of the environmental objectives and targets

- Findings of the EMS audits
- Evaluation of the effectiveness of the EMS
- Evaluation of the suitability of the environmental policy and the need for change in light of:
✓ Changing legislation
✓ Changing expectations and requirement of interested parties
✓ Changes in the products or activities of the organization
✓ Advances in science and technology
✓ Lessons learned from environmental accidents
✓ Market preferences
✓ Reporting and communication

Keeping in mind the commitment to continuous improvement, management should plan corrective and protective action to improve the EMS. In some cases, the management review may dictate changes in the environmental policy and thereby trigger changes in the EMS itself.

Non-Conformance, Corrective and Preventive Actions.

The findings, conclusions and recommendations reached from monitoring, audits and other reviews of the EMS should be documented and the necessary corrective and preventive actions identified. It is the task of the management to ensure that these corrective and preventive actions have been implemented and that there is a systematic follow-up to ensure their effectiveness. The main issue is not just to identify the problem, but to understand why it happened and to change the system, so that the same mistake does not occur twice. It is important to focus on the root-cause and not on the symptoms. The basic requirements include:

- Defining responsibility and authority for handling and investigating non-conformance
- Acting to mitigate the resulting impacts on the environment
- Initiating and completing corrective and preventive action
- Implementing and recording changes to documented procedures that result from corrective and preventive action.

Record keeping is very essential in EMS. The purpose of records is to demonstrate conformance to the requirements of the standards. In consequence, the procedures for identifying, maintaining and disposing of environmental records should be established. Environmental records include training records, audit records and reviews (ISO14001). Environmental records must be eligible, identifiable and traceable to the authority, product or service involved. The records must be protected against damage, deterioration or loss. Electronic environmental management systems could be created to keep track of the vast amounts of environmental data generated.

4.1.7 External Environmental Communication

This is a priority in environmental reporting and can be used for both internal and external communication. Environmental reports are rapidly becoming a primary channel for an organization's environmental communications. They are ways of providing documented evidence of the environmental challenges the organization faces and its efforts to manage them. In environmental reporting, three things are should be considered, namely, what needs to be said, to whom and how? Regular communication with external stakeholders is vital for attracting capital, maintaining a favorable public image, gaining market share and attracting and retaining the best employees, among other things. Environmental reporting has no rules on the frequency of disclosure, but EMAS and financial groups require disclosure on annual basis. Environmental and financial reports can supplement one another if developed at the same time. The content of an environmental report covers the following topics (EMAS):

- Environmental policy
- Environmental strategy
- Description of the EMS components
- Policy regarding environmental aspects related to products and services
- Listing of all inputs (materials and energy) and outputs (air pollution, sewage, waste) over the respective period of time (environmental effects register).
- Assessment of compliance with environmental regulations (environmental regulations register)
- Evaluation of environmental performance
- Description of the environmental management programs, including objectives, targets, measures and schedule
- Relationship with external stakeholders
- Audit report findings

Linking environmental reporting with financial reporting is not mentioned in any of the EMS standards, but it is important. It is a communication strategy that takes into account the interests of financial stakeholders like shareholders, banks, insurers and suppliers. Management must show that the chosen environmental strategy is efficient and effective and is creating shareholder value. An environmental strategy that does not create future free cash flow is economically unsustainable. The disclosure of environmental performance should ideally conform to the main qualitative criteria used in financial reporting, as exemplified below:

- Understandability – language should be simple, with technical matters banished to the appendices.
- Relevance – the information should be central to the user's decision-making, including confirming past evaluation or pending future behavior of the organization.
- Reliability – measurement and presentation should be free from material error or bias.
- Comparability – the information presented should be comparable over time and across organizations. This includes informing the user of methods used in measuring and presenting the information.

The structure of an environmental report should conform to the format and structure of a financial report since it is a supplementary document to the financial report. This facilitates reading and will make it easier to combine the two reports, once the financial and accounting standards are revised to incorporate environmental costs and benefits. An integration of both environmental and financial reports would give an added value to the organizational reporting system. A useful structure for this is given by Holmark (Holmark 1995, modified):

1. Contextual environmental issues
    i)   Type of industry
    ii)   Company profile and history
    iii)   Type of environmental effects of the industry and in the product life cycle

iv)   Other contextual environmental information

2. Environmental management report
   i)   Environmental policy
   ii)   Environmental objectives and targets
   iii)   Environmental programs
   iv)   Environmental management system
   v)   Environmental insurance

3. Environmental accounting policies
   i)   Definition of terms
   ii)   Financial environmental performance measurement principles
   iii)   Physical environmental performance measurement principles
   iv)   Legal environmental performance measurement principles

4. Financial environmental data and performance
   i)   Environmental costs
   ii)   Environmental benefits
   iii)   Environmental investments
   iv)   Environmental liabilities
   v)   Cost-benefit and risk assessment

5. Physical environmental performance (Environmental effects register and assessment)
   i)   Inputs
   ii)   Outputs
   iii)   Risk, incidents, corrective and preventive actions.

6. Legal environmental performance (Environmental regulations register and assessment)

i)   Relevant regulations

ii)   Changes in regulations and adoption by the organization

iii)   Compliance status (legal performance)

iv)   Non-conformance, corrective and preventive actions

7. Notes

8. Environmental verification statement.

4.2 Commitment to Performance Improvements

The direct benefits to an enterprise of implementing an EMS usually come from savings through cleaner production and waste minimization approaches. (An order of magnitude estimate is that about 50% of the pollution generated in a typical "uncontrolled" plant can be prevented, with minimal investment, by adopting simple and cheap process improvements.) Even in industrial countries, increased discharge fees and waste disposal charges provide incentives for cost-effective pollution reduction—which, incidentally, demonstrates the importance of an appropriate framework of regulations and incentives to drive the performance improvements. The major impact of the introduction of an EMS can be the identification of waste minimization and cleaner production possibilities. Management and worker commitment to improving performance is essential. The process of introducing the EMS can be a catalyst for generating support for environmental performance improvements, including the simple changes that make up "good housekeeping," and also for making the best use of existing pollution control equipment. Just as important, the development of good management systems is one of the best hopes for sustaining the improvements that can be achieved when attention is focused on environmental performance.

A concern often expressed about the ISO 14001 system is the lack of a clear commitment to improvements in actual environmental performance. The whole EMS approach is designed to improve performance, but critics of the rush to implement ISO 14001 argue that the standard can be misused. It is not yet clear how valid this point is, and its resolution will depend on how the overall approach is used in trade and regulatory areas.

However, there is a legitimate concern that some may view ISO 14000 as an end rather than a means.

Given the current stage of development of *auditing and certification systems*, it may be possible in some places to obtain (or claim) certification with a minimum level of real environmental improvement. From the World Bank's point of view, it is essential that enterprises demonstrate serious good-faith efforts to achieve the performance goals underlying an environmental management system, if certification is to have any real meaning. An acceptable system must comply with the spirit of the EMS, not just the minimum formal requirements.

4.3 Certification

ISO 14001 sets out a system that can be audited and certified. In many cases, it is the issue of certification that is critical or controversial and is at the heart of the discussion about the trade implications.

Certification means that a qualified body (an "accredited certifier") has inspected the EMS system that has been put in place and has made a formal declaration that the system is consistent with the requirements of ISO 14001. The standard allows for "self-certification," a declaration by an enterprise that it conforms to ISO 14001. There is considerable skepticism as to whether this approach would be widely accepted, especially when certification has legal or commercial consequences. At the same time, obtaining certification can entail significant costs, and there are issues relating to the international acceptance of national certification that may make it particularly difficult for companies in some countries to achieve credible certification at a reasonable cost. For firms concerned about having certification that carries real credibility, the costs of bringing in international auditors are typically quite high, partly because the number of internationally recognized firms of certifiers is limited at present. The issue of accreditation of certifiers is becoming increasingly important as the demand increases.

Countries that have adopted ISO 14001 as a national standard can accredit qualified companies as certifiers, and this will satisfy national legal or contractual requirements. However, the fundamental purpose of ISO is to achieve consistency internationally. If certificates from certain countries or agencies are not fully accepted or are regarded as "second class," the goal will not have been achieved. It is probable that the international marketplace will eventually put a real commercial value on high-quality certificates but this level of sophistication and discrimination has not yet been achieved. It is essential to the ultimate success of the whole system that there be a mechanism to ensure that certification in any one country has credibility and acceptability elsewhere. The ISO has outlined procedures for accreditation and certification (Guides 61 and 62), and a formal body, QSAR, has been established to manage the process. At the same time, a number of established national accreditation bodies heavily involved in ISO have set up the informal International Accreditation Forum (IAF) to examine mechanisms for achieving international reciprocity through multilateral agreements (MLAs). However, these systems are in the early stages, and many enterprises continue to use the established international certifiers, even at additional cost, because of lack of confidence in the acceptability of local certifiers. Given the variability in the design of individual EMS and the substantial costs of the ISO
14000 certification process, there is a growing tendency for large companies that are implementing EMS approaches to pause before taking this last step. After implementing an EMS and confirming that the enterprise is broadly in conformance with ISO 14001, it is becoming routine to carry out a "gap analysis" to determine exactly what further actions would be required to achieve certification and to examine the benefits and costs of bringing in third-party certifiers.

4.4 Trade Implications

Statements have been made to the effect that before long, ISO 14000 certification will be an essential passport for developing countries wishing to trade with the industrial nations. Such statements, in this extreme form, are speculative and almost certainly incorrect. It is, however, unclear to what extent ISO 14001 might become a barrier to trade, in direct contradiction to the basic objectives of the ISO, or, alternatively, might provide a competitive edge for certified firms. The trade implications are of concern to many countries, and the World Trade Organization (WTO) is beginning to consider some of the issues under its mandate on technical barriers to trade. In this context, a distinction needs to be made between product standards, such as the eco-labeling and LCA standards under ISO 14000, and production process standards such as ISO 14001; the impacts are likely to be different. In many cases in developing countries, the environmental pressures come through the supplier chain—the ongoing relationship between a major company (often a multinational) and its smaller national suppliers. The sensitivity of multinationals to pressures regarding their performance on environmental and other issues is causing them to look for better performance from the suppliers. This relationship is typically a cooperative one in which large companies work with smaller ones to achieve better performance in such areas as quality and price. The multinationals may ask their suppliers to achieve and demonstrate environmental performance improvements, but there is no evidence that unreasonable targets or time scales are being applied. Where ISO 14001 certification is an ultimate aim, certification is seen as a long-term objective rather than a short-term requirement. Even if ISO 14001 is not likely to be a contractual constraint in the foreseeable future, environmental performance is increasingly becoming a factor in commercial transactions, and companies looking to establish a presence in the international marketplace are considering whether a "green badge" would be an advantage to them. In practice, it is often marketing rather than environmental concerns that drive the ISO certification process.

4.5 Reducing the Cost of Regulation

A question commonly raised is the extent to which an EMS can reduce the costs of regulation, in terms of both the overall government enforcement effort and the costs of compliance of the individual enterprise. The use of ISO 14001 certification to replace some statutory reporting requirements is a topic of considerable discussion in a number of countries, particularly those where regulatory requirements are extensive enough to be a real burden on industry. It is now clear that an EMS is not a substitute for a regulatory framework, but the monitoring and reporting systems of a well-managed enterprise might substitute for some of the statutory inspections, audits, and reports normally required under government regulations. The issue is when and how the government can trust the capabilities and commitment of an enterprise to self-monitor its environmental performance and whether some formal EMS and certification system, such as ISO 14000, would provide the mechanism to convince regulators that scarce government resources would be better used elsewhere in pursuing less cooperative organizations. This approach is attractive, but there are a number of hurdles to clear before it can be put into place on a widespread basis. Reaching agreement on such matters is proving to be a more difficult and complex task than might at first be assumed. Some of the difficulties are legal (lack of flexibility in regulations or the need to ensure that voluntary reports are not unreasonably used to prosecute enterprises that are making genuine efforts to improve), but often they relate to the necessary level of confidence on both sides that the other parties are genuine in their efforts. Pilot programs being tested in a number of U.S. States will provide essential feedback on these issues. There are clear benefits all around in making such partnerships work, but it will be some time before clear, workable models are available.

4.6 Disclosure of Information and External Relations

There is considerable evidence that an informed public has a strong influence on the environmental performance of industrial enterprises, through a variety of mechanisms that include market forces, social pressures, and support for improved regulatory controls. ISO 14000 does not include specific requirements for the disclosure or publication of environmental performance measures or audit results, but other EMS models do have some such requirements. The World Bank and other environmental agencies and organizations strongly support disclosure of actual performance information because this allows the relevant public to monitor progress (or the lack of it) and to take informed positions on issues related to plant performance. It also allows much higher confidence in company statements about compliance and improvements. There is a growing interest on the part of commercial banks and insurance companies in environmental risk (in a purely business sense). Such organizations are considering whether EMS certification (typically EMAS, in Europe) demonstrates that a firm has real control over its environmental risk and potential liability. It is possible that certification may lead to commercial benefits, such as lower insurance rates, in certain high-risk sectors. Public release of the main environmental information from an EMS can also be used as a central component of a community relations program, although this goes beyond the basic concept of an EMS. Where social license to operate is at risk, release of the main environmental information from an EMS can add value to the transparency and accountability agenda of the organization.

4.7 Application to Small and Medium-Size Enterprises

Most of the development and application of EMS has taken place in large companies. The use of such systems in small and medium-size enterprises (SMEs) has been limited—although it is in this segment of industry that some of the largest benefits might be anticipated, because of the difficulty of regulating large numbers of small firms and the potential efficiency improvements that are believed to exist. In practice, however, the characteristics of the typical SME make the adoption of EMS difficult: most SMEs do not have a formal management structure, they lack technically trained personnel, and they are subject to severe short-term pressures on cash flow. It should be noted that an EMS cannot be used to drive improved performance in a poorly organized SME. Targeted training in management and quality control can improve overall performance, including its environmental aspects, and can provide a basis for more specific EMS development. Many firms can reap significant benefits from introducing quality management concepts, even where they are not aiming at formal certification. Any steps in this direction should be encouraged. EMS is not exclusive to large companies, it is for every organization. Every person is an environmental user but not every person is an environmental manager. Government had controlled the monopoly of environmental management for a very long time without achieving environmental sustainability. Now is the time for every person to be involved in environmental management, especially at an organized level like small and medium enterprises.

4.8 Role of Governments

Although ISO 14000 is a set of voluntary standards that individual companies may or may not choose to adopt, governments can clearly have a role in providing information, establishing the necessary framework and infrastructure, and, in some cases, helping companies to develop the basic capabilities to adopt ISO 14000. There are two particular areas in which government action would be useful: (a) providing information on the sectors and markets where ISO 14001 certification is a significant issue and assisting sector organizations to develop appropriate responses, and (b) helping to establish a certification framework, based on *strengthening national standards organizations* and encouraging competitive private sector provision of auditing and certification services.

Governments should see EMS approaches as part of a broad environmental strategy that includes regulatory systems, appropriate financial incentives, and encouragement of improved industrial performance. Such encouragement can really only be effective where there is cooperation at the government level between the relevant departments, including industry and trade, as well as environment. There is a growing interest in integrating environmental management issues into productivity or competitiveness centers designed to promote SME performance, but little information exists on experience to date.

CHAPTER FIVE

5.0 EMERGING AND CONTEMPORARY ISSUES

5.1 Challenges

EMS is clearly a good concept and is supported in principle by the World Bank and by environmental agencies and organizations everywhere. At the same time, there are costs associated with its implementation—particularly in enterprise time and effort, more than direct out-of-pocket costs— and a number of issues need to be addressed in making decisions about the type and level of system to be adopted. With the involvement of a relatively sizeable workforce (management and workers) in the implementation of the EMS, cost of training can be high and small organizations may not be able to afford it. The question of institutional capacity building of stakeholders on EMS is a critical factor. Lack of knowledge and skills on the part of communities has placed them on a disadvantaged position instead of a level playground. The voluntary nature of EMS tends to create confusion between it and corporate social responsibility and has become a source of tension and conflict development where negative environmental impacts are recorded. Emerging issues from this unhealthy interface has given rise to the incorporation of EMS concerns into memorandum of understanding agreements.

5.2 Practicalities in Establishing an EMS

The establishment of EMS builds on existing production and quality management systems.

Where such systems are weak or ineffective, as is often the case in enterprises that have poor environmental performance, a better management framework has to be established before focusing on the details of the EMS. The costs of establishing an EMS will therefore obviously depend on the starting point in terms of both management systems and environmental performance. The eco-efficiency savings can, in some cases, pay for the costs of establishing the EMS, particularly if most of the planning and organizational work is carried out in-house. However, a poor performer will very likely have to invest in production upgrading or pollution control in order to meet environmental requirements, and these costs can be significant. A full EMS can be complex and can require an appreciable commitment of operational resources. However, the final system can be reached reasonably through a series of discrete steps, starting from a basic, simple procedure and becoming more comprehensive and sophisticated as capabilities and resources allow. In this way, ***even a small enterprise can begin to put in place the basic elements of an ISO 14001 system and can develop them at an appropriate pace***. Once the basic EMS is in place, it is possible to carry out a gap analysis and to make a balanced judgment on the costs and benefits of seeking certification. A related issue is the coverage of the EMS. Certification is normally for specific sites or facilities. A large enterprise may have a number of different sites and production facilities and may choose to seek certification only for a subset of the sites.

5.3 Conclusion

The future prospects of ISO 14000 series is very promising. The spectacular blossoming of interest in ISO 14000 should lead to increased understanding of the benefits of better environmental management and greater awareness of environmental performance as a factor in succeeding in increasingly competitive markets. At the same time, this standard is not a magic wand that will achieve environmental improvements where regulation and enforcement are ineffective or that can open markets where competition is strong. Environmental governance cannot be achieved without good governance. In an atmosphere of endemic corruption and militancy, regulation and enforcement of standards cannot succeed. The standard provides a framework on which to build better performance, greater efficiency, and a competitive image. With serious commitment and effort from the organization, implementing a system such as ISO 14001 can yield solid benefits. It should be noted that sustainable development cannot thrive in an atmosphere of conflicts. Environmental issues are very sensitive and should be handled with care as emerging conflicts, if not well managed, could degenerate into crisis. In this era of globalization and increased awareness, it is obvious that future trends in business successes shall favor organizations whose processes and products are strongly rooted in the environmental management systems.

REFERENCES

Australian International Development Assistance Bureau. Appraisals, Evaluations and Sectoral Studies Branch. (1991). Environmental Assessment Guidelines for International Development Cooperation in the Agriculture Sector. Activity Guideline No. 2. Canberra, Australia: Australian Government Publishing Service.

Boyle, J. and Patterson, H. (Agrodev Canada Inc.). (2002). Environmental Sourcebook for Small-Scale Community Development Projects. Working Draft Prepared for CIDA, CIDA Internal Document (unpublished).

Canadian International Development Agency. (1995). Environmental Assessment at CIDA. Gatineau, Quebec: CIDA.

Canadian International Development Agency. (1996). Manual on the Canadian Environmental Assessment Act: The Canada Fund and Mission-Administered Funds. Gatineau, Quebec: CIDA.

Canadian International Development Agency. (2003) CEAA Work Tool v. 2. On-line version and off-line version. Gatineau, Quebec: CIDA.

Canadian International Development Agency. (2003). Desertification: a tool kit for programming. Version 1.0, 2003. Gatineau, Quebec: CIDA.

Chambers, R. (2002). Rural Appraisal: Rapid, Relaxed and Participatory. Brighton, United Kingdom: Institute of Development Studies, Discussion Paper No. 311.

Cressman, D.R., Zahedi, K. and Pinter, L. (2000). Capacity Building for Integrated Environmental Assessment and Reporting: Training

Manual (2nd edition). Winnipeg, Manitoba: International Institute for Sustainable Development (IISD) and United Nations Environment Programme (UNEP). Available at http://www.iisd.org/publications/publication.asp?pno=310

CSIR. Division of Water, Environment and Forest Technology. (1996). Strategic Environmental Assessment (SEA). A Primer. Stellenbosch, South Africa: CSIR. Available at http://fred.csir.co.za/www/sea/primer/primerf.htm

Dalal-Clayton, B. and Sadler, B. (1999). Strategic Environmental Assessment: A Rapidly Evolving Approach. Environmental Planning Issues No.18. London, United Kingdom: International Institute for Environment and Development. Available at http://www.nssd.net/pdf/IIED02.pdf do Rosário

Dearden, P., Jones, S. and Sartorius, R. (2002). Tools for development: A handbook for those engaged in development activity. London, United Kingdom: Department for International Development (DFID).

Dougherty, T.C. and Hall, A.W. (1995). Environmental impact assessment of irrigation and drainage projects. FAO Irrigation and Drainage Paper 53. London, United Kingdom: Overseas Development Administration of the UK (ODA); Rome, Italy: Food and Agriculture Organization of the United Nations (FAO).

DPR (1991). Department of Petroleum Resources, Environmental Guidelines and Standards for
the Petroleum Industry in Nigeria.

EU. (1992). Towards Sustainability: A European Programme of Policy and Action in relation to the Environment and Sustainable Development, Vol. II, 27 March 1992.

EU. (1996). Directive 96/61/EEC: Integrated Pollution Prevention and Control, Article 2(11).

FEPA (1989). Federal Environmental Protection Agency, National Policy on Environment
(Nigeria).

FEPA (1991). Federal Environmental Protection Agency, Guidelines and Standards for
Environmental Pollution Control in Nigeria.

FEPA (1992). Federal Environmental Protection Agency, Environmental Impact Assessment
Decree 86 of 1992 (Nigeria).

FEPA (1995). Federal Environmental Protection Agency Environmental Impact Assessment
Procedural Guidelines (Nigeria).

FEPA (1995). Federal Environmental Protection Agency Environmental Impact Assessment
Sectoral Guidelines for Oil and Gas Industry Projects (Nigeria).

FEPA (1999). Federal Environmental Protection Agency, National Guidelines for Environmental
Audit in Nigeria.

German Federal Ministry for Economic Cooperation and Development (BMZ). (1995). Environmental Handbook. Documentation on Monitoring and Evaluating Environmental Impacts. Volume I: Introduction, Crosssectoral Planning, Infrastructure. Volume II: Agriculture, Mining/Energy, Trade/Industry. Volume III: Compendium of Environmental Standards. Eschborn, Germany: Vieweg.

Government of Canada. (1999, 2004). Strategic Environmental Assessment: The Cabinet Directive on the Environmental Assessment of Policy, Plan and Program Proposals. Guidelines for Implementing the Cabinet Directive. Ottawa, Ontario: Government of Canada.

Guijt, I. (1998). Participatory monitoring and impact assessment of sustainable agriculture initiatives. SARL Discussion Paper No. 1. London, United Kingdom: International Institute for Environment and Development (IIED).

Holmark et al (1995). The Annual environmental Report: Measuring and reporting Environmental Performance. Copenhagen: Price Waterhouse.

ISO 10241. (1992). *International terminology standards — Preparation and layout*

ISO 14050. (2009). Environmental Management – Vocabulary

ISO/IEC Guide 2 (1996). *Standardization and related activities — General vocabulary*

Kelly, C. (2001). Rapid Environmental Impact Assessment: A Framework for Best Practice in Emergency Response. Disaster Management Working Paper 3/2001, Benfield Greig Hazard Research Centre. Presented at "Sharing Experiences on Environmental Management in Refugee Situations: A Practitioner's Workshop," Geneva, Switzerland, October 22–25, 2001.

Knausenberger, W.I., Booth, G.A., Bingham, C.S. and Gaudet, J.J. (1996). Environmental Guidelines for Small-Scale Activities in Africa. Environmentally Sound Design for Planning and Implementing Humanitarian and Development Activities. Washington, D.C., U.S.: U.S. Agency for International Development.

Available at http://www.encapafrica.org/resources.htm.

North American Commission for Environmental Cooperation. (2000). Guidance Document. Improving Environmental Performance and Compliance: 10 Elements of Effective Environmental Management Systems. Montreal, Quebec: North American Commission for Environmental Cooperation. Available at http://cec.org/files/PDF/LAWPOLICY/guide-e_EN.pdf

OECD. (1972). Council Recommendation on Environment and Economics: Guiding principles concerning international economic aspects of environmental policies.

OECD (1975). The Polluter Pays Principle.

Pallen, D. (1996). Environmental Assessment Manual for Community Development Projects. Gatineau, Quebec: CIDA, Asia Branch.

Pallen, D. (1997). Environmental Sourcebook for Micro-finance Institutions. Gatineau, Quebec: CIDA, Asia Branch.

Partidario, M., reviewed by Leblanc, P. and Fischer, K. (April 1996). Bibliography on Strategic Environmental Assessment. Ottawa, Ontario: Minister of Supply and Services Canada. Available at http://www.ceaa.gc.ca/017/0005/CEAA_4E.pdf

Schneider, H. and Libercier, M.-H. (1995). Participatory development: from advocacy to action. Paris, France: Organization for Economic Co-operation and Development.

Standards Organization of Nigeria (2002). ISO 14000: 1996 Environmental Management Systems Training Manual.

Standards Organization of Nigeria (2003). ISO 14001: 1996 Environmental Management Systems Training Manual.

Sturm, A. (1998) ISO 14001- Implementing an Environmental Management System (Version 2.02, 1998)

The President's Council on Sustainable Development. (1996). "Sustainable America: A New Consensus for Prosperity, Opportunity, and Health Environment for the Future":

UNEP/ICC/FIDIC. (1995). "Environmental Management System Training Resource Kit." Version 1.0.

United Nations Economic and Social Commission for Asia and the Pacific. (1990). Environmental Impact Assessment: Guidelines for Water Resources Development. ESCAP Environment and Development Series. New York, U.S.: United Nations.

United Nations High Commissioner for Refugees (UNHCR). (2002). Refugee Operations and Environmental Management. A Handbook of Selected Lessons Learned from the Field. Geneva, Switzerland: UNHCR.

Woodside, G., Yturri, J. and Aurricho, P. (1998). ISO 14001 Implementation Manual. New York, U.S.: McGraw-Hill Books.

World Bank. (1994). Environmental Assessment Sourcebook. Volume III. Guidelines for Environmental Assessment of Energy and Industry Projects. World Bank Technical Paper Number 154. Washington, D.C., U.S.: World Bank. Available at http://lnweb18.worldbank.org/ESSD/envext.nsf/47ByDocName/Too lsEnvironmentalAssessmentSourcebookandUpdates.

World Bank. (1995). Environmental Assessment Sourcebook. Volume II. Sectoral Guidelines. World Bank Technical Paper Number 140. Washington, D.C., U.S.: World Bank. Available at http://lnweb18.worldbank.org/ESSD/envext.nsf/47ByDocName/Too lsEnvironmentalAssessmentSource book and Updates

World Bank Group (1998), Implementing Policies: Industrial Pollution Management. Pollution Prevention and Abatement Handbook.

APPENDICES

APPENDIX A:

Environmental management —Vocabulary Management (ISO 14050)

This International Standard contains definitions of fundamental concepts related to environmental management, published in the ISO 14000 series of International Standards.

Terms and definitions
In some cases, the special usage of a concept in a particular context is indicated by the qualification given in angle brackets before the definition. The relevant source is given in brackets for each definition and note. If the same definition appears in more than one document, the earliest document is normally given as source.

1. General terms relating to environmental management

1.1
environment

surroundings in which an *organization* (1.4) operates, including air, water, land, natural resources,
flora, fauna, humans, and their interrelation

NOTE: Surroundings in this context extend from within an organization to the global system.

[ISO 14001]

1.2
environmental aspect

element of an *organization's* (1.4) activities, products or services that can interact with the *environment*(1.1)

NOTE: A significant environmental aspect is an environmental aspect that has or can have a significant *environmental impact* (1.3).

[ISO 14001]

1.3
environmental impact

any change to the *environment* (1.1), whether adverse or beneficial, wholly or partially resulting from an *organization's* (1.4) activities, products or services

[ISO 14001]

1.4
organization

company, corporation, firm, enterprise, authority or institution, or part or combination thereof, whether incorporated or not, public or private, that has its own functions and administration.

NOTE For organizations with more than one operating unit, a single operating unit may be defined as an organization.

[ISO 14001]

1.5
interested party

person or group having an interest in the performance or outcome of an organization or a system

NOTE 1 "Outcome" includes products and agreements; "system" includes product systems and environmental labeling and declaration systems.

NOTE 2 This generic definition is not taken directly from any other document. The concept is defined specifically from the point of view of environmental performance in ISO 14001 (with identical definition in ISO 14004 and ISO 14031), of type I environmental labelling in ISO 14024, of type III environmental declaration in ISO/TR 14025 and of life cycle assessment in ISO 14040.

The definitions are as follows:
□□□□individual or group concerned with or affected by the environmental performance of an organization
[ISO 14001]

□□□any party affected by a type I environmental labeling programme
[ISO 14024]

□□□any party affected by the development and use of a type III environmental declaration
[ISO/TR 14025]

□□□individual or group concerned with or affected by the environmental performance of a product system, or by the results of the life cycle assessment
[ISO 14040]

1.6
third party

person or body that is recognized as being independent of the parties involved, as concerns the issues in question

NOTE 1 "Parties involved" are usually supplier ("first party") and purchaser ("second party") interests.

[ISO 14024]

NOTE 2 "Third party" does not necessarily imply the involvement of a certification body.

[ISO/TR 14025]

1.7
certification

procedure by which a *third party* (1.6) gives written assurance that a product, process or service conforms to specified requirements

[ISO 14024]

1.8
prevention of pollution

use of processes, practices, materials or products that avoid, reduce or control pollution, which may include recycling, treatment, process changes, control mechanisms, efficient use of resources and material substitution

NOTE: The potential benefits of prevention of pollution include the reduction of adverse *environmental impacts*
(1.3), improved efficiency and reduced costs.

[ISO 14001]

1.9
waste

anything for which the generator or holder has no further use and which is discarded or is released to the
environment (1.1)

[ISO 14021]

NOTE "Waste" is also defined from the point of view of life cycle assessment in ISO 14040 as: "any output from the product system which is disposed of ".

1.10
transparency

open, comprehensive and understandable presentation of information

[ISO 14040]

1.11
environmental performance

results of an *organization's* (1.4) management of its *environmental aspects* (1.2)

NOTE In the context of **environmental management systems** (2.1), results may be measured against the **organization**'s (1.4) **environmental policy** (2.1.1), **objectives** (2.1.2), and **targets** (2.1.3).

[ISO 14031]

1.11.1
environmental performance evaluation (EPE)

process to facilitate management decisions regarding an *organization's* (1.4) *environmental performance* (1.11) by selecting indicators, collecting and analyzing data, assessing information against *environmental performance criteria* (1.11.1.1), reporting and communicating,
and periodically reviewing and improving this process

[ISO 14031]

1.11.1.1
environmental performance criterion

environmental objective (2.1.2), *target* (2.1.3), or other intended level of *environmental performance* (1.11) set by the management of the *organization* (1.4) and used for the purpose of *environmental performance evaluation* (1.11.1)

[ISO 14031]

1.11.1.2
environmental condition indicator (ECI)

specific expression that provides information about the local, regional, national, or global condition of the *environment* (1.1)

NOTE "Regional" may refer to a state, a province, or a group of states within a country, or it may refer to a group of countries or a continent, depending on the scale of the condition of the environment that the **organization** (1.4) chooses to consider.

[ISO 14031]

1.11.1.3
environmental performance indicator (EPI)

specific expression that provides information about an *organization's* (1.4) *environmental performance* (1.11)

[ISO 14031]

1.11.1.3.1
management performance indicator (MPI)

environmental performance indicator (1.11.1.3) that provides information about the management efforts to influence an *organization's* (1.4) *environmental performance* (1.11)

[ISO 14031]

1.11.1.3.2
operational performance indicator (OPI)

environmental performance indicator (1.11.1.3) that provides information about the *environmental performance* (1.11) of an *organization's* (1.4) operations

[ISO 14031]

2. Terms relating to environmental management systems

2.1
environmental management system (EMS)

part of the overall management system that includes organizational structure, planning activities, responsibilities, practices, procedures, processes and resources for developing, implementing, achieving, reviewing and maintaining the *environmental policy* (2.1.1)

[ISO 14001]

2.1.1
environmental policy

statement by the *organization* (1.4) of its intentions and principles in relation to its overall *environmental performance* (2.1.5) which provides a framework for action and for the setting of its *environmental objectives* (2.1.2) and *targets* (2.1.3)

[ISO 14001]

2.1.2
environmental objective

overall environmental goal, arising from the *environmental policy* (2.1.1), that an *organization* (1.4) sets itself to achieve, and which is quantified where practicable

[ISO 14001]

2.1.3
environmental target

detailed performance requirement, quantified where practicable, applicable to the *organization* (1.4) or parts thereof, that arises from the *environmental objectives* (2.1.2) and that needs to be set and met in order to achieve those objectives

[ISO 14001]

2.1.4
continual improvement

process of enhancing the *environmental management system* (2.1) to achieve improvements in overall *environmental performance* (2.1.5) in line with the *organization's* (1.4) *environmental policy* (2.1.1)

NOTE: The process need not take place in all areas of activity simultaneously.

[ISO 14001]

2.1.5
environmental performance

☐management system☐ ☐measurable results of the *environmental management system* (2.1), related to an *organization's* (1.4) control of its *environmental aspects* (1.2), based on its *environmental policy*
(2.1.1), *objectives* (2.1.2) and *targets* (2.1.3)

[ISO 14001]

3. Terms relating to auditing

3.1
environmental audit

systematic, documented verification process of objectively obtaining and evaluating *audit evidence* (3.4) to determine whether specified environmental activities, events, conditions, management systems, or information about these matters conform with *audit criteria* (3.3), and communicating the results of this process to the client

[ISO 14010]

3.1.1
environmental management system audit

systematic and documented verification process of objectively obtaining and evaluating *audit evidence*
(3.4) to determine whether an *organization's* (1.4) *environmental management system* (2.1) conforms
with the environmental management system audit criteria, and communicating the results of this process to the client

[ISO 14011]

3.1.2
environmental management system audit

☐internal audit☐ ☐systematic and documented verification process of objectively obtaining and evaluating evidence to determine whether an *organization's* (1.4) *environmental management system* (2.1) conforms to the environmental management system *audit criteria* (3.3) set by the organization, and for
communication of the results of this process to management

[ISO 14001]

3.2
subject matter

specified environmental activity, event, condition, management system, and/or information about these matters

[ISO 14010]

3.3
audit criteria

policies, practices, procedures or requirements against which the auditor compares collected *audit evidence* (3.4) about the subject matter

NOTE: Requirements may include but are not limited to standards, guidelines, specified organizational requirements, and legislative or regulatory requirements

3.4
audit evidence

verifiable information, records or statements of fact

NOTE 1: Audit evidence, which can be qualitative or quantitative, is used by the auditor to determine whether **audit criteria** (3.3) are met.

NOTE 2: Audit evidence is typically based on interviews, examination of documents, observation of activities and conditions, existing results of measurements and tests or other means within the scope of the audit.

[ISO 14010]

3.5
audit finding

result of the evaluation of the collected *audit evidence* (3.4) compared against the agreed *audit criteria* (3.3)

NOTE The findings provide the basis for the audit report.

[ISO 14010]

3.6
audit conclusion

professional judgement or opinion expressed by an auditor about the *subject matter* (3.2) of the audit, based on and limited to reasoning the auditor has applied to *audit findings* (3.5)

[ISO 14010]

3.7
auditee

organization (1.4) to be audited

[ISO 14010]

3.8
audit client

client organization (1.4) commissioning the audit

NOTE 1: The client may be the **auditee** (3.7), or any other organization which has the regulatory or contractual right to commission an audit.

[ISO 14010]

NOTE 2: In ISO 14010, the term "client" is used instead of audit client.

3.9
audit team

group of auditors, or a single auditor, designated to perform a given audit

NOTE 1: The audit team may also include technical experts and auditors-in-training.

NOTE 2: One of the auditors on the audit team performs the function of lead auditor.

NOTE 3: Adapted from ISO 14010.

3.9.1
environmental auditor

person qualified to perform *environmental audits* (3.1)

[ISO 14010]

3.9.2
lead environmental auditor

person qualified to manage and perform *environmental audits* (3.1)

[ISO 14010]

3.9.3
technical expert

☐auditing☐ ☐person who provides specific knowledge or expertise to the *audit team* (3.9), but who does not
participate as an auditor

[ISO 14010]

4. Terms relating to product system

4.1
product system

collection of materially and energetically connected *unit processes* (4.3) which perform one or more defined functions

NOTE 1: For the purposes of life cycle assessment, the term "product" used alone includes not only product systems but can also include service systems.

NOTE 2: Adapted from ISO 14040.

4.2
product

any goods or service

[ISO 14021]

4.2.1
intermediate product

input (4.12) to or *output* (4.13) from a *unit process* (4.3) which requires further transformation

[ISO 14041]

4.2.2
co-product

any of two or more products from the same *unit process* (4.3)

[ISO 14041]

4.2.3
packaging

material that is used to protect or contain a *product* (4.2) during transportation, storage, marketing or use

NOTE 1: For the purposes of type II environmental labelling, the term "packaging" also includes any item that is physically attached to, or included with, a product or its container for the purpose of marketing the product or communicating information about the product.

NOTE 2: Adapted from ISO 14021.

4.2.4
final product

product (4.2) which requires no additional transformation prior to its use

[ISO 14041]

4.3
unit process

smallest portion of a *product system* (4.1) for which data are collected when performing a *life cycle assessment* (5.3)

[ISO 14040]

4.4
functional unit

quantified performance of a *product system* (4.1) for use as a reference unit in a *life cycle assessment* (5.3) study

[ISO 14040]

4.5
system boundary

interface between a *product system* (4.1) and the *environment* (1.1) or other product systems

[ISO 14040]

4.6
allocation

partitioning the *input* (4.12) or *output* (4.13) flows of a *unit process* (4.3) to the *product system* (4.1) under study

[ISO 14040]

4.7
elementary flow

☐input☐ ☐material or energy entering the system being studied, which has been drawn from the *environment* (1.1) without previous human transformation

NOTE Adapted from ISO 14040.

4.8
elementary flow

□output□ □material or energy leaving the system being studied, which is discarded into the **environment** (1.1) without subsequent human transformation

NOTE: Adapted from ISO 14040.

4.9
raw material

primary or secondary material that is used to produce a **product** (4.2)

[ISO 14040]

4.10
energy flow

input (4.12) to or **output** (4.13) from a **unit process** (4.3) or **product system** (4.1), quantified in energy
units

NOTE: Energy flow that is input may be called energy input; energy flow that is output may be called energy output.

[ISO 14041]

4.10.1
feedstock energy

heat of combustion of raw material inputs, which are not used as an energy source, to a **product system**
(4.1)

NOTE: It is expressed in terms of higher heating value or lower heating value.

[ISO 14041]

4.10.2
process energy

energy input required for a *unit process* (4.3) to operate the process or equipment within the process, excluding energy inputs for production and delivery of this energy

[ISO 14041]

4.11
reference flow

measure of the needed *outputs* (4.13) from processes in a given *product system* (4.1) required to fulfill the function expressed by the *functional unit* (4.4)

[ISO 14041]

4.12
input

material or energy which enters a *unit process* (4.3)

NOTE Materials may include *raw materials* (4.9) and *products* (4.2).

[ISO 14040]

4.12.1
ancillary input

material input that is used by the *unit process* (4.3) producing the product, but does not constitute a part of the product

EXAMPLE : A catalyst.

[ISO 14041]

4.13
output

material or energy which leaves a *unit process* (4.3)

NOTE: Materials may include **raw materials** (4.9), **intermediate products** (4.2.1), **products** (4.2), emissions and **waste** (1.9).

[ISO 14040]

4.13.1
fugitive emission

uncontrolled emission to air, water or land

EXAMPLE: Material released from a pipeline coupling.

[ISO 14041]

5. Terms relating to life cycle assessment

5.1
life cycle

consecutive and interlinked stages of a *product system* (4.1), from raw material acquisition or generation of natural resources to the final disposal

[ISO 14040]

5.2
practitioner

individual or group that conducts a *life cycle assessment* (5.3)

[ISO 14040]

5.3
life cycle assessment (LCA)

compilation and evaluation of the *inputs* (4.12), *outputs* (4.13) and the potential *environmental impacts* (1.3) of a *product system* (4.1) throughout its *life cycle* (5.1)

[ISO 14040]

5.3.1
life cycle inventory analysis

phase of *life cycle assessment* (5.3) involving them compilation and quantification of *inputs* (4.12) and
outputs (4.13), for a given *product system* (4.1) throughout its *life cycle* (5.1)

[ISO 14040]

5.3.1.1
life cycle inventory result (LCI result)

outcome of a *life cycle inventory analysis* (5.3.1) that includes flows crossing the *system boundary*
(4.5) and provides the starting point for *life cycle impact assessment* (5.3.2)

[ISO 14042]

5.3.1.2
data quality

characteristic of data that bears on their ability to satisfy stated requirements

[ISO 14041]

5.3.1.3
uncertainty analysis

systematic procedure to ascertain and quantify the uncertainty introduced into the results of a *life cycle inventory analysis* (5.3.1), due to the cumulative effects of input uncertainty and data variability

NOTE: Either ranges or probability distributions are used to determine the uncertainty in the results.

[ISO 14041]

5.3.1.4
sensitivity analysis

systematic procedure for estimating the effects on the outcome of a study of the chosen methods and data

[ISO 14041]

5.3.2
life cycle impact assessment (LCIA)

phase of *life cycle assessment* (5.3) aimed at understanding and evaluating the magnitude and significance of the potential *environmental impacts* (1.3) of a *product system* (4.1)

[ISO 14040]

5.3.2.1
impact category

class representing environmental issues of concern into which *LCI results* (5.3.1.1) may be assigned

[ISO 14042]

5.3.2.1.1
life cycle impact category indicator

quantifiable representation of an *impact category* (5.3.2.1)

[ISO 14042]

5.3.2.2
characterization factor

factor derived from a model which is applied to convert *LCI results* (5.3.1.1) to the common unit of the *life cycle impact category indicator* (5.3.2.1.1)

[ISO 14042]

5.3.2.3
environmental mechanism

system of physical, chemical and biological processes for a given *impact category* (5.3.2.1), linking *LCI results* (5.3.1.1) to *category indicators* (5.3.2.1.1) and *category endpoints* (5.3.2.4)

[ISO 14042]

5.3.2.4
category endpoint

attribute or aspect of natural environment, human health or resources, identifying an environmental issue of concern

[ISO 14042]

5.3.3
life cycle interpretation

phase of *life cycle assessment* (5.3) in which the findings of either the inventory analysis or the impact
assessment, or both, are combined consistent with the defined goal and scope in order to reach conclusions and recommendations

[ISO 14040]

5.3.3.1
consistency check

process of verifying that the assumptions, methods and data are consistently applied throughout the study and in accordance with the goal and scope definition

NOTE The consistency check should be performed before conclusions are reached.

[ISO 14043]

5.3.3.2
sensitivity check

process of verifying that information obtained from a *sensitivity analysis* (5.3.1.4) is relevant for reaching the conclusions and giving recommendations

[ISO 14043]

5.3.3.3
completeness check

process of verifying whether information from the preceding phases of an *life cycle assessment* (5.3) or *life cycle inventory analysis* (5.3.1) is sufficient for reaching conclusions in accordance with the goal and scope definition

[ISO 14043]

5.3.4
comparative assertion

environmental claim regarding the superiority or equivalence of one product versus a competing product which performs the same function

[ISO 14040]

6. Terms relating to environmental labelling and declarations

6.1
environmental claim

statement, symbol or graphic that indicates an *environmental aspect* (1.2) of a *product* (4.2), a component or *packaging* (4.2.3)

NOTE: An environmental claim may be made on product or packaging labels, through product literature, technical bulletins, advertising, publicity, telemarketing, as well as through digital or electronic media such as the Internet.

[ISO 14021]

6.1.1
environmental label

environmental declaration claim which indicates the *environmental aspects* (1.2) of a *product* (4.2) or service

[ISO 14020]

6.1.2
qualified environmental claim

environmental claim (6.1) which is accompanied by an *explanatory statement* (6.1.4) that describes the limits of the claim

[ISO 14021]

6.1.3
environmental claim verification

confirmation of the validity of an ***environmental claim*** (6.1) using specific predetermined criteria and procedures with assurance of data reliability

[ISO 14021]

6.1.4
explanatory statement

any explanation which is needed or given so that an ***environmental claim*** (6.1) can be properly understood by a purchaser, potential purchaser or user of the ***product*** (4.2)

[ISO 14021]

6.2
type I environmental labelling programme

voluntary, multiple-criteria-based, third party programme that awards a licence which authorizes the use of ***environmental labels*** (6.1.1) on ***products*** (4.2) indicating overall environmental preferability of a
product within a particular ***product category*** (6.2.1) based on ***life cycle*** (5.1) considerations

[ISO 14024]

6.2.1
product category

group of *products* (4.2) which have equivalent function

[ISO 14024]

6.2.1.1
fitness for purpose

ability of a product, process or service to serve a defined purpose under specific conditions

[ISO 14024]

6.2.1.2
product function characteristic

attribute or characteristic in the performance and use of a *product* (4.2)

[ISO 14024]

6.2.1.3
product environmental criteria

environmental requirements that the *product* (4.2) shall meet in order to be awarded an *environmental label* (6.1.1)

[ISO 14024]

6.2.2
eco-labelling body

third party body, and its agents, which conducts a *type I environmental labelling programme* (6.2)

[ISO 14024]
6.3
self-declared environmental claim

environmental claim (6.1) that is made, without independent third-party certification, by manufacturers, importers, distributors, retailers or anyone else likely to benefit from such a claim

[ISO 14021]

NOTE This is also called "type II environmental labelling".

6.4
type III environmental declaration

quantified environmental data for a product with preset categories of parameters based on the ISO 14040
series of standards, but not excluding additional environmental information provided with a type III environmental declaration programme

[ISO/TR 14025]

6.4.1
type III environmental declaration programme

voluntary process by which an industrial sector or independent body develops a type III environmental declaration, including setting minimum requirements, selecting categories of parameters, defining the involvement of third parties and the format of external communications

[ISO/TR 14025]

6.5
upgradability

characteristic of a ***product*** (4.2) that allows its modules or parts to be separately upgraded or replaced without having to replace the entire product

[ISO 14021]

6.6
material identification

words, numbers or symbols used to designate composition of components of a ***product*** (4.2) or ***packaging*** (4.2.3)

NOTE: A material identification symbol is not considered to be an **environmental claim** (6.1).

[ISO 14021]

APPENDIX B (informative)

Additional terms and definitions from Technical Report (ISO/TR 14061)

B.1
forest

generally considered to be a plant community of predominantly trees and other woody vegetation growing
together, its land, flora and fauna, their interrelationships, and the resources and values attributed to it

NOTE: Forests vary greatly around the world depending on the climate, soil, history and culture of the country involved. Many countries have a definition of forest included in legislation.

B.2
principles, criteria and indicators

international, national and private sector initiatives, whether governmental or non-governmental, provide a common hierarchical framework including "Principles, criteria and indicators" for evaluating progress towards achieving SFM

NOTE 1: In some initiatives, the principles are considered to be included in the criteria.

NOTE 2: For the purposes of this report, the term "Criteria & Indicators" is used specifically in reference to the sets of Criteria & Indicators of Sustainable Forest Management developed through the intergovernmental processes.

B.3
principles

fundamental rules which serve as a basis for reasoning and action

NOTE Principles are explicit elements of a goal such as SFM.

B.4
criteria

characteristics that are considered important and by which success or failure can be judged

NOTE: The role of criteria is to characterize or define the essential elements or set of conditions or processes by which sustainable forest management may be assessed.

[Source: Intergovernmental Seminar on Criteria and Indicators for SFM (ISCI)]

B.5
indicators

quantitative, qualitative or descriptive measures that when periodically evaluated and monitored show the direction of change

[Source: Intergovernmental Seminar on Criteria and Indicators for SFM (ISCI)]

B.6
sustainable development

meeting the needs of the present without compromising the ability of future generations to meet their own
needs

[Source: The Brundtland Report]

B.7
sustainable forest management

NOTE: While there is broad agreement on the concept of SFM, there are variations in the definitions developed through the various national and international initiatives. Two definitions of SFM have been included here so that the user of this Technical Report can understand the scope of the concept and the ways it has been defined by people from two different regions of the world.

B.7.1
sustainable forest management (SFM)
process of managing permanent forest land to achieve one or more clearly specified objectives of management with regard to the production of a continuous flow of desired forest products and services, without undue reduction of its inherent values and future productivity and without undue undesirable effects on the physical and social environment

[Source: International Tropical Timber Organization (ITTO)]

B.7.2
sustainable forest management (SFM)

stewardship and use of **forests** (A.1) and forest land in a way and at a rate that maintains their biodiversity, productivity, regeneration capacity, vitality and their potential to fulfil, now and in the future, relevant ecological, economic and social functions, at local, national and global levels and does not cause damage to other ecosystems

[Source: Pan-European (Helsinki) Process]

APPENDIX C (informative)

Additional concepts encountered in the international environmental community

C.1 Best Available Technique (BAT)

[1] EU Directive 96/61/EEC (September 24, 1996) concerning Integrated Pollution Prevention
and Control, Article 2(11).

[2] OECD Council Recommendation, May 1972, Environment and Economics, Guiding principles
concerning international economic aspects of environmental policies.

[3] Convention on the Protection of the Marine Environment of the North East Atlantic. Paris 22 September 1992, Article 2, clause 3 (b) and amendment No. 1.

C.2 Critical load

[1] DOWING, R.J., HETTELINGH, J.-P. and DE SMET, P.A.M., 1993. Calculation and Mapping Critical Loads in Europe. Status Report 1993.

C.3 Precautionary principle

[1] ISO 14004:1996, *Environmental management systems — General guidelines on principles, systems and supporting techniques*, Annex A, principle No. 15.

[2] The Rio Declaration on Environment and Development, principle No. 15.

[3] Convention on the Protection of the Marine Environment of the North East Atlantic. Paris, 22 September 1992. Article 2, clause 2 (a).

C.4 "Polluter pays" principle

[1] ISO 14004:1996, *Environmental management systems — General guidelines on principles, systems and supporting techniques*, Annex A, principle No. 16.

[2] The Rio Declaration on Environment and Development, principle No. 16.

[3] Convention on the Protection of the Marine Environment of the North East Atlantic. Paris, 22 September 1992. Article 2, clause 2 (b).

[4] The Polluter Pays Principle, OECD 1975.

C.5 Pollution

[1] EU Directive 96/61/EEC (September 24, 1996) concerning Integrated Pollution Prevention and Control, Article 2(11).

[2] IMO/UNESCO/WMO/IAEA/UN/UNEP Joint Group of experts on the Scientific Aspects of Marine Pollution (GESAMP).

[3] Convention on the Protection of the Marine Environment in the North East Atlantic. Paris, 22 September 1992. Article 1, clause (d).

[4] Convention on the Protection of the Marine Environment of the Baltic Sea Area, 1992,
(Helsinki Convention), Article 2, clause 1.

C.6 Sustainable development

[1] "Our Common Future": Report published by the World Commission on Environment and the Development (the Brundtland Report).

[2] "Sustainable America: A New Consensus for Prosperity, Opportunity, and Health Environment for the Future": The President's Council on Sustainable Development, February 1996.

[3] Towards Sustainability: A European Programme of Policy and Action in relation to the Environment and Sustainable Development. EU, Vol. II, 27 March 1992.

ALPHABETICAL INDEX

A
allocation 4.6
ancillary input 4.12.1
audit client 3.8
audit conclusion 3.6
audit criteria 3.3
audit evidence 3.4
audit finding 3.5
audit team 3.9
auditee 3.7

B
BAT B.1 **best available technique** B.1

C
category endpoint 5.3.2.4
category indicator 5.3.2.1.1
certification 1.7
characterization factor 5.3.2.2
client 3.8
comparative assertion 5.3.4
completeness check 5.3.3.3
consistency check 5.3.3.1
continual improvement 2.1.4
co-product 4.2.2
criteria A.4
critical load B.2

D
data quality 5.3.1.2

E

ECI 1.11.1.2
ecolabelling body 6.2.2
elementary flow 4.7, 4.8
EMS 2.1
energy flow 4.10
environment 1.1
environmental aspect 1.2
environmental audit 3.1
environmental auditor 3.9.1
environmental claim 6.1
environmental claim verification 6.1.3
environmental condition indicator 1.11.1.2
environmental declaration 6.1.1
environmental impact 1.3
environmental label 6.1.1
environmental management system 2.1
environmental management system audit 3.1.1
environmental management system audit 3.1.2
environmental mechanism 5.3.2.3
environmental objective 2.1.2
environmental performance 1.11
environmental performance 2.1.5
environmental performance criterion 1.11.1.1
environmental performance evaluation 1.11.1
environmental performance indicator 1.11.1.3
environmental policy 2.1.1
environmental target 2.1.3
EPE 1.11.1
EPI 1.11.1.3
explanatory statement 6.1.4

F
feedstock energy 4.10.1
final product 4.2.4
fitness for purpose 6.2.1.1
forest A.1
fugitive emission 4.13.1
functional unit 4.4

I
impact category 5.3.2.1
indicators A.5
input 4.12
interested party 1.5
intermediate product 4.2.1

L
LCA 5.3
LCIA 5.3.2
LCI results 5.3.1.1
lead environmental auditor 3.9.2
life cycle 5.1
life cycle assessment 5.3
life cycle impact assessment 5.3.2
**life cycle impact category
indicator** 5.3.2.1.1
life cycle interpretation 5.3.3
life cycle inventory analysis 5.3.1
life cycle inventory result 5.3.1.1

M
**management performance
indicator** 1.11.1.3.1
material identification 6.6
MPI 1.11.1.3.1

O
**operational performance
indicator** 1.11.1.3.2
OPI 1.11.1.3.2
organization 1.4
output 4.13

P
packaging 4.2.3
"polluter pays" principle B.4

Printed in Great
Britain
by Amazon

31619960R00071